T0305781

ERGONOMICS
Foundational Principles, Applications, and Technologies

Ergonomics Design and Management: Theory and Applications

Series Editor
Waldemar Karwowski
Industrial Engineering and Management Systems
University of Central Florida (UCF) – Orlando, Florida

Published Titles

Ergonomics: Foundational Principles, Applications, and Technologies
Pamela McCauley Bush

Aircraft Interior Comfort and Design
Peter Vink and Klaus Brauer

Ergonomics and Psychology: Developments in Theory and Practice
Olexiy Ya Chebykin, Gregory Z. Bedny, and Waldemar Karwowski

Ergonomics in Developing Regions: Needs and Applications
Patricia A. Scott

Handbook of Human Factors in Consumer Product Design, 2 vol. set
Waldemar Karwowski, Marcelo M. Soares, and Neville A. Stanton

> Volume I: Methods and Techniques
> Volume II: Uses and Applications

Human–Computer Interaction and Operators' Performance: Optimizing Work
Design with Activity Theory
Gregory Z. Bedny and Waldemar Karwowski

Trust Management in Virtual Organizations: A Human Factors Perspective
Wiesław M. Grudzewski, Irena K. Hejduk, Anna Sankowska, and Monika Wańtuchowicz

Forthcoming Titles

Knowledge Service Engineering Handbook
Jussi Kantola and Waldemar Karwowski

Manual Lifting: A Guide to the Study of Simple and Complex Lifting Tasks
Daniela Colombiani, Enrico Ochipinti, Enrique Alvarez-Casado, and Thomas R. Waters

Neuroadaptive Systems: Theory and Applications
Magalena Fafrowicz, Tadeusz Marek, Waldemar Karwowski, and Dylan Schmorrow

Organizational Resource Management: Theories, Methodologies, and Applications
Jussi Kantola

ERGONOMICS
Foundational Principles,
Applications, and Technologies

Pamela McCauley Bush, PhD,CPE

CRC Press
Taylor & Francis Group
Boca Raton London New York

CRC Press is an imprint of the
Taylor & Francis Group, an **informa** business

CRC Press
Taylor & Francis Group
6000 Broken Sound Parkway NW, Suite 300
Boca Raton, FL 33487-2742

© 2012 by Taylor & Francis Group, LLC
CRC Press is an imprint of Taylor & Francis Group, an Informa business

No claim to original U.S. Government works

ISBN-13: 978-1-4398-0445-2 (hbk)

Library of Congress Cataloging-in-Publication Data

McCauley-Bush, Pamela.
    Ergonomics : foundational principles, applications, and technologies / Pamela McCauley-Bush.
        p. cm. -- (Ergonomics design & management theory & applications)
    "A CRC title."
    Includes bibliographical references and index.
    ISBN 978-1-4398-0445-2 (alk. paper)
    1. Human engineering--Textbooks. I. Title.

TA166.M33 2012
620.8'2--dc23   2011039818

Visit the Taylor & Francis Web site at
http://www.taylorandfrancis.com

and the CRC Press Web site at
http://www.crcpress.com

This book is dedicated to my parents, Maurice Sr. and LaFrance McCauley. Daddy, thank you for making me take Algebra in the 7th grade, which began my love for math, science, and ultimately engineering. Mom, thanks for your eternal faith in me, constant encouragement, and steadfast optimism.

Contents

Preface

The motivation for this book is a result of almost two decades of teaching ergonomics and human factors courses. In my interaction with students, it was apparent to me that an introductory text that focuses on basic ergonomic principles from both a research and application perspective could be very useful. This book is written to be used at the undergraduate and graduate levels as well as by professionals in industry requiring a first course in ergonomics. It is designed to be used as that introductory text and accompanied with hands-on laboratory activities to complement classroom instruction. The use of case studies is designed to demonstrate application of the ergonomic knowledge. The book covers the foundational principles and major topics in the field of physical ergonomics. Upon completion of a course using this book, the student will be prepared to apply the ergonomic knowledge in industry or continue to higher levels of study in the field.

The content covered in the book includes the following:

- Chapter 1: Introduction to ergonomics, history of the field, and associated legislative agencies.
- Chapter 2: Overview of the systems body; this will serve as a refresher of select material learned in anatomy, biology, and physiology courses.
- Chapter 3: Overview of human sensory capabilities as well as measurement of environmental factors.
- Chapter 4: Introduction to muscular work and innervations of the muscular system.
- Chapter 5: Introduction to the study of human body dimensions (or anthropometry).
- Chapter 6: Student application of ergonomic principles to the design of workplaces and hand tools.
- Chapter 7: Introduction to work-related musculoskeletal disorders, including history of disorders, causal factors, and mitigation strategies.
- Chapter 8: Introduction to work physiology with a discussion of the principles of heavy work, and evaluating physical workload and fatigue.
- Chapter 9: Overview of cognitive ergonomics and information processing as well as an explanation of principles for the design of controls and displays.
- Chapter 10: Introduction to product liability and the impact of ergonomics in design on the consumer.

DESCRIPTION OF LABORATORY MANUAL

The laboratory manual contains multiple laboratory and application assignments to give students hands-on experience in applying ergonomic material taught in the classroom lectures. The manual has labs for each of the primary topics covered in the course as well as guidelines on how students are to conduct the laboratories and prepare lab report. Numerous tables, equations and examples are provided in the lab manual to facilitate student understanding of the material. The use of the lab manual supports the instructor by providing tailored exercises for students to perform that are directly aligned with the textbook material. Assignments are also provided for students taking the course via distance learning or remote resources. http://www.crcpress.com/product/isbn/9781439804452

Acknowledgments

The completion of this project is the result of tremendous support by many people in my professional and personal life. I want to thank CRC Press for the opportunity to be a part of this exciting series, specifically editor Cindy Carelli. I would also like to thank my mentors, Dr. Adedeji Badiru, Dr. Howard Adams, and Dr. Deborah Reinhart, for their persistent support and guidance, as well as for serving as outstanding examples of successful academic authors.

I am forever grateful to many wonderful students I have had the opportunity to teach at the University of Central Florida and the Massachusetts Institute of Technology, as their desire to learn ergonomics and human factors has been a primary motivator for me. I want to thank Dr. Rochelle Jones, Dr. Samiullah Durrani, Johnine Mowatt, Juanita Salazar, Christina Rusnock, Shanon Wooden, and Adrienne Brown for their assistance with the research. A special thank you to my PhD student, Larry Lowe, for the many hours he spent collecting material and resources. Rachelle Borelli, thank you for the creative and insightful introductions you prepared for the chapters. I want to also thank my sister, Pipina Figaro Smith, for support in the editing. A special acknowledgment goes to my brilliant graduate student, Susan Gaines, for her commitment and exemplary support in the final stages of this project—you are amazing. To my former assistant, Angela Shirley, I say thank you for your outstanding organizational skills and commitment. To my fantastic assistant Deborah LaClair, this project could not have been completed without your attention to detail, tireless commitment, and outstanding support; Deborah, words cannot begin to express how much I appreciate your effort and excellence in the completion of this project.

On a personal note, I want to thank my loving and supportive family. My daughter and son-in-law, Annette and Nate Hemphill, thank you for the constant words of encouragement and Sunday brunch when I desperately needed it! I want to thank my sisters, Princess Hill, LaWanna Porter, and Pipina Figaro Smith, for always believing in me. To my big brother, Maurice McCauley, Jr., thanks for the quiet encouragement and inspiration only you can give. To my wonderful prayer partner, Sandra Jeter, thank you for the many prayers, affirmations, and laughter. To my parents, Maurice and LaFrance McCauley, thanks for giving me the vision to be an engineer and teach. Finally, to my wonderful husband, Michael Bush, thank you for seeing gifts in me that I may not see in myself. Your strength, patience, encouragement, and support in keeping me focused throughout this project made the difference. You all have my greatest love and appreciation.

Sincerely,

Pamela McCauley Bush, PhD, CPE

Author

Dr. Pamela McCauley Bush is an associate professor in the Department of Industrial Engineering and Management Systems at the University of Central Florida, where she leads the Human Factors in Disaster Management Research Team. Her research focus includes human factors in disaster management, evaluation and development of artificial intelligence models using fuzzy set theory, human factors model development, human factors in chemical and biological weapon development, and the human impact in information security. She is the author of over 60 technical papers, book chapters, and conference proceedings. Dr. Bush serves as a member of the editorial board for the journal of *Theoretical Issues in Ergonomics Sciences* (TIES), is the associate editor of the *Industrial Engineering Encyclopedia*, and referees technical research papers for *IEEE Transactions on Systems, Man, and Cybernetics*, the *International Journal of Industrial Hygienist*, the *North American Fuzzy Information Processing Conference Proceedings*, and Kluwer Academic Publishers. In 2006, her research group was awarded the Best Paper Award in Human Factors at the Annual Industrial Engineering Research Conference. Dr. Bush has been described as an outstanding professor and teacher. Her teaching efforts have resulted in the receipt of both the College of Engineering Award for Excellence in Undergraduate Teaching and the Teaching Incentive Program Award (TIP). To inspire and support those walking the paths she has taken, Dr. Bush is an active advisor and mentor. Her track record of working closely with minority and female engineering students has resulted in over 75% of her female and minority protégés obtaining advanced degrees. Her dedication to making a difference for underrepresented students is evident in the fact that over 50% of the students that she serves as PhD advisor to are African American and 40% are female. Additionally, she previously served as the faculty advisor to the University of Central Florida (UCF) Chapter of the National Society of Black Engineers and is advisor for the UCF Chapter of the Society of Women Engineers. In addition to her commitment to the university, Dr. Bush is the chief technology officer (CTO) of Bush Enterprises; the parent company of Antone & Associates (a consulting firm); Technology Solutions & Innovations. The company is led by her husband Michael A. Bush, chief executive officer. Bush Enterprises provides engineering support, consulting, and software development services. Clients have included the Department of Defense, NASA, universities, and corporate customers. A native of Oklahoma, Dr. Bush received her bachelor's, master's, and doctor of philosophy degrees in industrial engineering from the University of Oklahoma. In fact, in 1993, Dr. Bush became the first known African-American female to be granted a PhD in the field of engineering in the state of Oklahoma. In 2005, she graduated from the Advanced Minority Business Executive Program (AMBEP) at the Dartmouth College Tuck School of Business.

Additionally, Dr. Bush has the distinction of being a certified professional ergonomist (CPE). She has held various leadership positions in the business, technology, and education communities. She previously served on the Orange County Florida Board of Zoning Adjustment at the appointment of the Mayor of Orange County, Florida, and was elected member of the board of directors for the National Center for Simulation. In 2004, she was selected to serve on the Blue Ribbon Panel for Orange County Public Schools. In 2001, she was appointed by Governor Jeb Bush's office to serve on the Florida Research Consortium, a position that she still holds today. Additionally, she has served on the National Board of Directors for the Women in Engineering Program Advocates Network (WEPAN) from 1998 to 2000. She presently maintains board membership in a number of agencies, including the Beta Parent Program, the University of Oklahoma Industrial Engineering Advisory Board, and the University of Oklahoma, College of Engineering Minority Engineering Advisory Board. Over the past 20 years, Dr. Bush has received numerous awards in recognition of her commitment, professional accomplishments, and community outreach efforts. One of her proudest moment was when she recently received the Distinguished Alumni Award from the University of Oklahoma. In 2007, she received the Engineer of the Year Award from the Florida Engineering Society and in 2006, was recognized by the Society of Women Engineers as Engineering Educator of the Year. She also received the Woman of Distinction in Technology Award from the Central Florida Girl Scout Council in 2006. She has been recognized as one of the 10 Small Business Women of the Year in Central Florida and the Millennium Woman of the Year 2001 by the Millennium Woman Foundation, Los Angeles, California. Other honors include the *Saturn/Glamour Magazine* "Women Making a Difference Award" 2000 and the Outstanding Woman of Color in Technology Award for Educational Leadership in 1999. Finally, Dr. Bush is a nationally recognized motivational speaker and author. She and her daughter, Annette Hemphill, produced *The Word on the Go!* CD series: a collection of scriptures designed to provide spiritual enrichment through hearing and applying the word of God. She is the author of *Winners Don't Quit ... Today They Call Me Doctor* and *The Winners Don't Quit Kit*, an autobiographical novel and empowerment package that inspires success academically, professionally, spiritually, and personally. As a public speaker and lecturer, she travels around the country to give inspirational and informative lectures on women's leadership issues, technology in America, and motivational topics. In these speaking events, she often shares her challenges and triumphs in business, education, and leadership. Additionally, she shares the benefits of determination, faith in God, and diligence in reaching for and achieving one's goals despite financial difficulties, disappointments, and being a teen mother. Her activities have led to recognition in national publications, including *Pink Magazine, Black Enterprise*, the *Orlando Business Journal, Lifestyle Magazine, Charisma Magazine*, the *Florida Engineering Society Magazine, Career Engineer, Ebony, Essence, Jet, Lear's Magazine*, the

Institute of Industrial Engineering Magazine, and *US Black Engineer*. She has also appeared on the Big Idea Show with Donny Deutsch, an internationally viewed program on CNBC. Dr. Bush is the proud wife of Michael A. Bush and together they have three daughters, Annette, Brandi, and Brittany, as well as a son-in-law, Jonathan. She is the daughter of Maurice Sr. and LaFrance McCauley of Oklahoma City, Oklahoma.

1 Foundational Ergonomics

1.1 LEARNING GOALS

This chapter will provide the student with a basic understanding of ergonomics, its definitions, and its applications. The student will also learn about the historical foundations of ergonomics by studying the individuals and organizations that led the early development of the field of ergonomics. Additionally, topics relevant to American ergonomic legislation and worldwide professional societies will be discussed.

1.2 KEY TOPICS

- Definition of ergonomics
- History of ergonomics
- Types of ergonomic problems
- Legislative and regulatory organizations impacting ergonomics
- Ergonomic societies and resources

1.3 INTRODUCTION AND BACKGROUND

"Ergonomics" is derived from the Greek words *ergo* (work) and *nomos* (laws) to denote the science of work. Ergonomics is a scientific discipline focused on comprehensively addressing the interaction of humans with all aspects of their environment. While ergonomics initially is focused on humans within the occupational setting, its wide purview and relevance to everyday human life has extended ergonomic application to cover other areas, including consumer product design, recreational environments, and processes.

The principles of ergonomics are not limited to traditional occupational environments such as offices or factories, but are broadly extended to address the needs of users' environments including the service, health care, and recreational industries. The objective of the application of ergonomic principles to any environment is to design the environment to be compatible with the need of the human users. In other words, ergonomic applications are designed to "fit the task to the person." This textbook is dedicated to teaching the physical ergonomic principles, tools, and techniques for the successful application of ergonomics in practice and research. The International Ergonomics Association (IEA) offers the following definition for ergonomics:

> Ergonomics (or human factors) is the scientific discipline concerned with the understanding of interactions among humans and other elements of a system, and the profession that applies theory, principles, data and methods to design in order to optimize human well-being and overall system performance. (IEA, 2009)

Frederick Taylor (1856–1915), often referred to as the "father of modern ergonomics," was an early pioneer in the field of industrial engineering. During his lifetime, Taylor's principles, often termed "scientific management" or "Taylorism," were considered extremely useful (Taylor, 1911). However, today his methods are seen in a negative light, similar to Fordism and other related assembly-line forms of production. These connotations, however, have more to do with the working conditions at the time than Taylor's studies. Having worked his way up from a machine shop laborer to the position of chief engineer, Taylor held the interests of the worker as highly as the interests of the owner, although his studies were often misused by owners to get the most work out of their employees for the lowest possible wages (Winkel and Westgaard, 1996).

Two of Taylor's disciples in the early 1900s were Lillian and Frank Gilbreth. The Gilbreth's 12 children contributed largely to their interest in management theory, and they developed, tested, and implemented efficiency methods in their own household in order to care for so many kids. One of their children later wrote a book called *Cheaper by the Dozen*, based on their family life which demonstrated the broad application of these principles in all areas of the Gilbreths' lives. The history of scientific applications to humans in the occupational environment by the Gilbreths provided a foundation for applied ergonomics and human factors.

1.4 ERGONOMICS DEFINED

This basic definition of ergonomics is broad and comprehensive, with the intent of addressing the occupational needs of humans in a holistic fashion. This perspective is shared among applied ergonomists and researchers. Although various definitions of ergonomics exist, the essence of each of these descriptions lies in applying scientific principles to design processes, systems, equipment, and environments to be compatible with the needs of the given population.

Various definitions of the field of ergonomics and human factors exist from the scientific literature, professional organizations, government agencies, and open sources. These definitions are largely consistent but lean toward the needs of the organization. Three definitions from the Human Factors and Ergonomics Society (HFES) are provided in the following text.

1.4.1 HUMAN FACTORS AND ERGONOMICS SOCIETY

The HFES is dedicated to the betterment of humankind through the scientific inquiry into and application of those principles that relate to the interface of humans with their natural, residential, recreational, and vocational environments and the procedures, practices, and design considerations that increase a human's performance and safety at those interfaces (HFES, 2011).

1.4.2 DICTIONARY OF HUMAN FACTORS AND ERGONOMICS

Human Factors is that field which is involved in conducting research regarding human psychological, social, physical, and biological characteristics, maintaining the information obtained from that research, and working to apply that information with respect to the design, operation, or use of products or systems for optimizing human performance, health, safety, and/or habitability. (HFES, 2011; Stramler, 1993)

1.4.3 NATIONAL AERONAUTICS AND SPACE ADMINISTRATION

National Aeronautics and Space Administration (NASA) defines human factors as follows:

> Human factors is an umbrella term for several areas of research that include human performance, technology design, and human-computer interaction. The study of human factors in the Human Factors Research and Technology Division at NASA Ames Research Center focuses on the need for safe, efficient and cost-effective operations, maintenance and training, both in flight and on the ground. (HFES, 2011; NASA AMES)

1.4.4 CETENA (THE ITALIAN SHIP RESEARCH COMPANY)

"Ergonomics is the study of human performance and its application to the design of technological systems". The goal of this activity is to enhance productivity, safety, convenience, and quality of life. Example topics include models and theories of human performance, design and analytical methodology, human–computer interface issues, environmental and work design, and physical and mental workload assessment. Human factors engineering requires input from disciplines ranging from psychology and environmental medicine to statistics. (This definition is found at http://www.cetena.it/ergostoria.htm) (HFES, 2011).

1.4.5 ERGONOMICS AND SAFETY

While preventing injuries and fatalities is essential to optimizing human well-being, safety is just one of the many aspects of ergonomics. As can be seen from IEA's definition, ergonomics also seeks to optimize overall performance. Ergonomics fits the task to the worker, not just to prevent injury, but also to increase productivity. Safety is intertwined with physical ergonomics topics such as defining optimal work postures and workstation layouts, minimizing repetitive motions, preventing musculoskeletal disorders, and facilitating manual material handling. However, ergonomics also includes cognitive and organizational topics, which expand beyond the limited bounds of safety, to include topics such as easing mental workload, improving human–computer interaction, optimizing policies and procedures, and ensuring quality.

1.4.6 ERGONOMICS DOMAINS

The ergonomics discipline can be classified into three domains to categorize the general area of emphasis. These domains are broadly categorized by the Federation of European Ergonomics Societies (FEES) as physical, cognitive, and organizational ergonomics. The FEES definitions of these areas are as follows (FEES, 2009):

- *Physical ergonomics* is concerned with human anatomical, anthropometric, physiological, and biomechanical characteristics as they relate to physical activity. Relevant topics include
  - Working postures
  - Materials handling

- Repetitive movements
- Work-related musculoskeletal disorders
- Workplace layout and design
- Safety and health

- *Cognitive ergonomics* is concerned with mental processes, such as perception, memory, reasoning, and motor response, with regard to interactions among humans and other elements of a system. Relevant topics include
  - Mental workload
  - Decision making
  - Skilled performance
  - Human–computer interaction
  - Human reliability
  - Stress
  - Training
- *Organizational ergonomics* is concerned with the optimization of socio-technical systems, including their organizational structures, policies, and processes. Relevant topics include
  - Communication
  - Personnel resource management
  - Task design
  - Design of shift hours
  - Team and cooperative work
  - Participatory design
  - Virtual organizations
  - Production
  - Quality management

These categories of ergonomics can be used as guidelines for assessing risk factors and mitigating the impact of these issues on workers in the occupational environment.

1.4.7 INTERDISCIPLINARY NATURE OF ERGONOMICS

Ergonomics is an interdisciplinary field where knowledge and theory are integrated to obtain understanding, practical tools, and technologies to address human needs in product design for the occupational environment and other diverse environments. Contributing disciplines in ergonomics include physiology, psychology, biomechanics (sometimes considered a subset of ergonomics), physics, anthropometry, and general engineering (Figure 1.1).

- *Physiology*—Understanding how the physical aspects of the human respond to the work environment.
- *Psychology*—Understanding the cognitive aspects of human interaction in the work environment.
- *Biomechanics*—Concerned with the mechanical elements of living organisms. Occupational biomechanics deals with the mechanical and

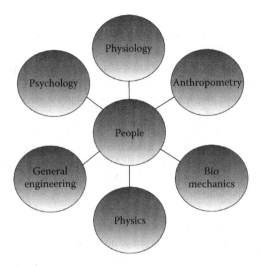

FIGURE 1.1   Interdisciplinary fields within ergonomics.

motion characteristics of the human body and its elements in the work environment.

- *Physics*—Uses laws of science and engineering concepts to describe motion undergone by the various body segments and the forces acting on these body parts during normal daily activities and job tasks.
- *Anthropometry*—The empirical science that attempts to define reliable physical measures of a person's size, form and capabilities for anthropological comparison.
- *General Engineering*—Used to develop appropriate tools, facility, and equipment designs. These foundations can be used to help classify the types of ergonomic problems that generally occur.

1.4.8   CLASSIFICATION OF ERGONOMIC PROBLEMS

In the evaluation of ergonomic problems, identification of the types of risk factors is useful in categorizing problems as they relate to the human. This user-based approach to design assessment allows processes and equipment to be evaluated based upon tasks requirements and users' abilities. The broad categories for ergonomic problems include general and sometimes overlapping areas. These areas are as follows:

1. Anthropometric
   a. Anthropometry is the study of human body dimensions and comprises the measurement of body proportions, including weight and volume. Reach distances, sitting eye height, and buttock-to-knee height are several examples. Dynamic capabilities, joint angle, and strength norms are also considered in anthropometrics.
   b. Applied anthropometrics applications can reduce the dimensional conflict between functional space geometry and the human body.

2. Musculoskeletal
    a. Tasks that strain the muscular and skeletal systems are in this category.
    b. Overexertion in this area can lead to injuries such as cumulative trauma disorders, back injuries, muscular strain, and muscle sprain.
3. Cardiovascular
    a. Cardiovascular problems arise in tasks that place stress on the heart and circulatory system.
    b. When a demand is placed on the cardiovascular system, the heart pumps more blood to the muscles to meet the elevated oxygen requirements. If the activity is excessive, it can lead to premature fatigue, overexertion, dehydration, and produce both long-term and short-term negative effects on the worker.
4. Cognitive
    a. Cognitive problems arise when there is either information overload or under-load in information-processing requirements.
        i. Both the short-term and the long-term memory may be strained in a task.
        ii. On the other hand, these functions should be sufficiently utilized for the maintenance of an optimum state of arousal.
        iii. Tasks should be designed to be compatible with cognitive capabilities and to complement human functions with machine functions for enhanced performance.
5. Psychomotor
    a. Psychomotor problems are those that strain the systems that respond to visual or auditory stimuli with a movement or reply.
    b. Paced manipulative work with significant visual demands is an example of such a task.

In an ergonomic assessment, if problems have been classified using the user-based classifications described earlier, a systematic approach to addressing the risk factors can then be applied. This systematic approach considers the category and identifies compatible assessment tools (e.g., musculoskeletal problems are identified in a lifting task, thus a lifting analysis will be performed to evaluate the task). This approach can be coupled with additional evaluation methods, such as those discussed in the following, to address ergonomic problems or used exclusively.

An alternative method to address ergonomic problems is a task based approach. This method includes understanding the area of study with respect to the occupational environment as the basis for evaluation. This approach focuses on the specific type of occupational issue and the resultant impact on the user in order to design ergonomic interventions or applications. The primary areas of ergonomics applications, as functions of a task based approach, include the following:

- Physical aspects of human–machine interaction
    - Size, shape, color, and texture of displays and controls for cars, domestic appliances, industrial and commercial equipment, etc.
    - Method of operation or interaction with the displays and controls

- Cognitive aspects if the human–machine interface and team interaction
  - Understanding of instructions and warnings
  - Style of dialogue between computer and user
  - Study of team dynamics including communication tools and team hierarchy
- Workplace design and workplace layout
  - Layout and design of processes, offices, factories, and equipment
  - All physical aspects of a task including hand tools, seating, and anything the worker interacts with to complete task requirements
- Physical environmental factors
  - Effects of noise, vibration, illumination, and chemical/biological contaminants on human performance and health
- Job design, selection, and training
  - Effects of job design, work schedules (i.e., shift work), worker selection process, instructions, administrative controls, training, and related factors on task performance
- Organizational environment or psychosocial factors
  - Organizational structure within a group and its effects on satisfaction with the task, productivity, and employee perception of the work environment

The use of this evaluation method can support a systems approach to the application of ergonomic solutions. Ideally, a hybrid of both approaches should be used. This results in a user-based perspective while considering all analysis aspects of occupational environment to produce a comprehensive approach.

1.4.9 History of the Field

Some of the earliest known indications of ergonomics are attributed to the Italian physician and philosopher Bernardino Ramazzini. He is considered the founder of occupational medicine and his findings made an early case for the field of ergonomics. In the 1700s, Ramazzini published *De morbis artificum diatriba* (Diseases of Workers, 2009), the first comprehensive work on occupational diseases (Diseases of Workers). Ramazzini, through his observations, revealed a variety of common workers' diseases that appeared to be caused by prolonged, irregular motions and postures during task performance. These musculoskeletal injuries were very similar to the soft tissue injuries (i.e., carpal tunnel syndrome or tendonitis) that workers experience today in the presence of excessive repetition, force, excessive joint deviation, and other known risk factors (McCauley-Bell, 1993).

The term "ergonomics" was first coined in 1857 by Wojciech Jastrzebowski, an author and Polish scientist (Helander, 2005). A few years later, in 1883 (Koppes, 2006), engineer Frederick Taylor would apply ergonomic principles to improve productivity in a factory setting by designing aspects of the work environment to be suitable for users. The application of ergonomic principles is as much about effective management of the work environment as it is about the science of the field. According to Frederick Taylor, the father of scientific management, management's main goal should be to maximize prosperity for the employer and employee. He sought maximum efficiency by focusing on the training and development of each worker, so that each individual produced the maximum output for his abilities. He argued that increasing efficiency not only benefited the employer through increased output, but also benefited the employee by enabling the employer to pay higher wages while still experiencing increased profit. In order to achieve this maximum efficiency, he called for management to (Koppes, 2006) do the following:

- Develop a science for each element of work to replace previously accepted general methods
- Scientifically select, train, teach, and develop each worker
- Cooperate with workers to ensure work is done according to the developed science
- Have management assume responsibility for tasks for which it is best suited, rather than have the worker perform all of the tasks

Taylor argued that by implementing these four fundamental principles, management can take on the burden of planning and applying scientific principles, leaving the worker to perform the tasks for which they are most suited. The result is increased prosperity for employers and improved job satisfaction for employees.

The formal development of the field of ergonomics began in the early 1900s and was largely linked to the changing industrial environment in the United States. Ergonomic science was further propelled by military needs for equipment to effectively meet the requirements of soldiers during combat. In war time, military scientists and psychologists began conducting research on humans in occupational settings. While much of this early ergonomic research would actually be classified today as "fitting the worker to the task" rather than "fitting the task to the worker," this work laid much of the foundation upon which the human factors and ergonomic field is now based.

In the early twentieth century (Price, 1989), Frank and Lillian Gilbreth conducted motion analysis studies that gave insight into the required movements and their associated times for a variety of common occupational actions. These became known as "therbligs." This term is the name "Gilbreth" spelled backward. The Gilbreths' work included the study of skilled performance and fatigue, as well as the design of workstations and equipment for users with disabilities (Wood and Wood, 2003). A summary of the major periods of ergonomic growth over the past 50 years are characterized in Table 1.1.

This timeline provides a broad overview of the history and focus areas of significant periods in ergonomics history. A more detailed timeline is shown at the end of the chapter which provides more insight into the history and growth of the field.

1.4.10 FEDERAL AGENCIES AND ERGONOMICS

Numerous U.S. government agencies have sponsored research or applications of human factors and ergonomics. Some of these agencies and examples of applications and research in ergonomics are listed below:

TABLE 1.1

Periods of Ergonomic Growth and Specific Ergonomic Events

Year	Activity
1700	Bernardino Ramazzini publishes *De Morbis Artificum Diatriba* (Diseases of Workers)
1857	Wojciech Jastrzebowski, a Polish scholar, first coins the term "ergonomics"
1883	Fredrick Taylor applies ergonomics principles to improve factory productivity
1900–1920	Frank and Lillian Gilbreth begin their work in motion study and shop management
1945–1960	Ergonomics begins to flourish in post–World War II societies; "knob and dial" era is born
1949	Ergonomics Research Society (now known as Ergonomics Society) is established in Britain
1957	The race for space and staffed space flight begins; human factors becomes an important part of the space program
1957	The Human Factors Society (now known as the Human Factors and Ergonomics Society) is founded
1957	The IEA is formed
1960–1980	Rapid growth and expansion of human factors in the United States
1960	Membership in the Human Factors Society reaches 500
1970	OSHA and NIOSH are both created by the U.S. Congress
1979	Three Mile Island incident has a major impact on the perception of a need for human factors in the workplace
1980	Membership in the Human Factors Society reaches 3000
1980–2010	Human factors continues to be an area of rapid growth and is also becoming a household name with commercials stating "ergonomically designed" as a buzz word
2001	OSHA issues an ergonomics standard, which requires companies to address hazards likely to cause sprains, strains, and repetitive motion injuries

Source: Salvendy, G., Handbook of human factors and ergonomics, John Wiley, 2006.

- *Department of Defense (DoD)*—Extensive applications and research in human factors and ergonomics to support the design of aircraft cockpits, weapons systems, processes, team interaction, and many other systems.
- *Federal Highway Administration*—Applications in the design of highways, road signage, and predicating driver behavior.
- *National Aeronautical Space Administration*—Applications to design for human capabilities and limitations in space, design of space stations and vehicles. NASA research has also studied how to design for effective human use in support in tasks on the earth such as maintenance.
- *National Highway Traffic Safety Administration*—Applications which involve the design of cars, driver distraction studies, and effects of drugs and alcohol on driving.
- *Department of the Interior*—Applications include ergonomics in numerous federally regulated fields such as underground mining and fire fighting.
- *National Institute of Standards and Technology*—Applications involve safe design of consumer products.

- *National Institute for Occupational Safety and Health*—Studies focus on ergonomic injuries in the workplace, industrial safety, and work stress.
- *Occupational Safety and Health Administration (OSHA)*—Provides private industries with guidelines, regulations, and standards for safety in the occupational environment.
- *Nuclear Regulatory Commission*—Involved in the design requirements for nuclear power plants. Examples of applications of human factors and ergonomics include design of controls and displays, warning systems, and monitoring systems.
- *Federal Aviation Administration (FAA)*—Deals with aviation safety. Widespread examples of applications of ergonomics are found in the design of equipment, processes, and tasks for workers from air traffic controllers to pilots.

These are a few instances of U.S. government agencies which rely on the use of ergonomic principles. The needs of these agencies have been useful in promoting research, study, and application of ergonomic principles. The success of these applications has led to growth in the field.

1.4.11 ERGONOMICS AND HUMAN FACTORS PROFESSIONAL ORGANIZATIONS

There are over 16,000 ergonomists working worldwide in research, application, educational, and training professions (APA, 2009). The educational background and the level of education vary among these professionals. General areas of study for ergonomic professionals may include industrial engineering, psychology, industrial hygiene, kinesiology, biomechanics, or related fields.

The advantage of joining a society is that it provides a forum for the exchange of ideas and support for the individuals working within the ergonomics field. Professional societies also work toward developing certifications and standards to maintain a level of excellence for its members. These organizations also provide a forum for continuing education and opportunities for members to stay current on new developments in the field.

Several ergonomic societies are located within the United States the UK and around the world. To be recognized as legitimate ergonomics associations, these societies usually chose to become members of the IEA. The IEA is a global federation of ergonomic societies from around the world. The IEA was established by the European Productivity Agency in 1959 with the objective "to elaborate and advance ergonomics science and practice, and to improve the quality of life by expanding its scope of application and contribution to society" (IEA, 2009).

The HFES is the largest national ergonomics professional organization with about 5000 members. HFES is organized into 23 technical interest groups and has several local and student chapters. The society also publishes three journals: *Human Factors: The Journal of the Human Factors and Ergonomics Society*, *Journal of Cognitive Engineering and Decision Making*, and *Ergonomics in Design: The Quarterly of Human Factors Applications*.

HFES defined its mission statement as follows:

The Human Factors and Ergonomics Society is dedicated to the betterment of human-kind through the scientific inquiry into and application of those principles that relate to the interface of humans with their natural, residential, recreational, and vocational environments and the procedures, practices, and design considerations that increase a human's performance and safety at those interfaces. (HFES website, 2009)

The Institute of Ergonomics and Human Factors (IEHF), formerly known as the Ergonomic Society, is located in the United Kingdom (http://www.ergonomics.org.uk/). This society was founded in 1950, and it is the oldest formal ergonomics organization in the world. It has around 800 members, mostly in the United Kingdom. The society supports two journals: *Ergonomics* and *Applied Ergonomics*. This society has been instrumental in the promotion and establishment of high standards for application, publication, and research in the field of ergonomics in Europe and worldwide.

1.4.12 LEGISLATIVE AND REGULATORY ISSUES IMPACTING ERGONOMICS

In the United States, the National Institute for Occupational Safety and Health (NIOSH) and the OSHA were both established by the Occupational Safety and Health Act of 1970 (OSHA Act, 1970, 2011). Both agencies seek safe and healthy working conditions; with NIOSH focused on research, information, education, and training, while OSHA focuses on developing and enforcing workplace regulations. To these ends, NIOSH serves as part of the Centers for Disease Control and Prevention (CDC) under the Department of Health and Human Services, while OSHA is a part of the U.S. Department of Labor (National Institute for Occupational Safety and Health, 2009). These government agencies have been tasked with providing legislation, guidelines, and tools to support ergonomic applications and safety in the occupational environment. OSHA also is responsible for promulgation of standards. Similar agencies exist in other countries such as

- Australia (Australian Safety and Compensation Council [ASCC])
- Canada (Canadian Centre or Occupational Health and Safety)
- South Korea (Korean Occupational Safety and Health Agency [KOSHA]), and many others

In addition to the governmental influence, numerous private organizations have contributed to the available guidelines and standards.

The primary legislative guidance in ergonomics and human factors in the United States is obtained from the OSHA; however, there are other agencies that influence the occupational environment as well as industries or governmental areas over which OSHA does not have authority. The other agencies or private organization impacting occupational safety legislation and guidelines include

- Department of Defense
  - Military Standard (Mil-STD) 1472F, 1999
- Environmental Protection Agency

BOX 1.1   MISSION AND IMPACT OF OSHA

Mission and Impact of OSHA

Mission: To prevent work-related injuries, illnesses, and deaths.

Impact: Since the agency was crated in 1971, occupational deaths have been cut by 62% and injuries have declined by 42%.

OSHA website (2009)

- Mining Enforcement Agency
- International Organization for Standardization (ISO)
  - ISO9000 series

A brief discussion of OSHA, NIOSH, and other legislative on standards related organizations follows.

1.4.12.1   OSHA

OSHA became the leading force in establishing and promulgating standards to promote safety and health in the occupational environment in the United States. The OSHA Act gives the agency the authority to (OSHA Act, 1970) do the following;

- Establish a set of criteria that the employer will use in protecting employees against health hazards and harmful materials.
  - Grant the Secretary of Labor, through OSHA, the authority, among other things, to do the following: promulgate, modify, and revoke safety and health standards.
  - Conduct inspections and investigations.
  - Issue citations, including proposed penalties.
  - Require employers to keep records of safety and health data.
  - Petition courts to restrain imminent danger situations.
  - Approve or reject state plans for programs under the act.

Though OSHA resources do not appear to be in proportion to the size of the American workforce, the presence of this organization has had a significant positive impact on the workplace and worker safety. The reduction in work-related deaths and injuries suggest that some of the approaches over the last few decades have, in fact, been effective at accomplishing the mission of OSHA.

Although no ergonomic standards exist, OSHA currently provides ergonomic-related guidelines for shipyards, poultry processing, meatpacking plants, retail grocery stores, and nursing homes. OSHA also provides a number of electronic tools for identifying and resolving hazards for numerous occupations including electrical contractors, baggage handling, beverage delivery, computer work, grocery warehousing, health care, poultry processing, printing, and sewing. OSHA's ergonomic outreach and assistance programs facilitated the development of programs that have become industry standards. These programs include the Alliance Program, Onsite Consultation Program, Safety and Health Achievement Recognition Program

(SHARP), Strategic Partnerships, and the Voluntary Protection Program (VPP). Additional analysis and risk assessment tools made available by NIOSH and OSHA directly or through funded research projects include the NIOSH Lifting Equation for Manual Material Handling, Liberty Mutual Tables' Push/Pull Analysis, Hand-Arm Vibration Analysis, and the Rapid Entire Body Assessment (REBA) Tool (Occupational Safety and Health Administration, 2011).

As previously mentioned, OSHA's mission includes not only development, but also enforcement of safe and healthy working conditions. Although OSHA does not currently operate under a specific ergonomic standard, OSHA can issue citations for ergonomic hazards under the General Duty Clause, Section 5(a) (1), which requires that each employer keep their workplaces free from recognized hazards that could likely cause death or serious physical harm (OSHA Act, 1970).

1.4.12.2 OSHA Standards

The OSHA Act is the culmination of guidelines and safety information that had been compiled by various industries and government agencies. These guidelines were from existing federal standards and national consensus standards such as the American National Standards Institute (ANSI) and the National Fire Protection Agency (NFPA). The guidelines offered by ANSI, NFPA, and numerous other organizations became the basis for many current occupational standards. The guidelines, norms, and standards of these national organizations were adapted with some amendments, deletions, and additions in order to be accepted as standards in the early days of OSHA.

The development of standards is a continuing process and is an ongoing effort at OSHA. These standards are developed in four categories including

* *Design standards*—Detailed design criteria for the construction of certain items or components. Design standards tend to be the most specific standards. An example is ventilation design details.
* *Performance standards*—States the objective that must be obtained and leaves the method for achieving it up to the employer. Such employer-specific standards are the threshold limit values (TLVs) for toxic and hazardous substances.
* *Vertical standards*—Apply to a particular industry with specifications that relate to individual operations. Examples of industry-specific standards include operation in the pulp, paper, and paperboard industries.
* *Horizontal standards*—Applies to all workplaces and relates to broad areas. These standards include those which can apply to any industry including those as diverse as sanitation or walking and working surfaces.

1.4.12.3 National Institute for Occupational Safety and Health (NIOSH)

The NIOSH was also established by the OSH Act of 1970. NIOSH was established to undertake health studies of alleged hazardous conditions, and to develop criteria to support revisions of, or recommendations to, OSHA for new health standards. NIOSH provides information and data about health hazards, although the final authority for promulgation of the standards remains with OSHA (Box 1.2) (NIOSH, 2009).

BOX 1.2   NIOSH MISSION

NIOSH Mission
The mission of NIOSH is
to generate new
knowledge in the field of
occupational safety and
health and to transfer
the knowledge into
practice for the
betterment of workers.

NIOSH website (2009)

To accomplish the stated mission, NIOSH does the following:

NIOSH provides national and world leadership to prevent work-related illness, injury, disability, and death by gathering information, conducting scientific research, and translating the knowledge gained into products and services, including scientific information products, training videos, and recommendations for improving safety and health in the workplace. (NIOSH website, 2011)

In keeping with the mission of NIOSH, the agency established the National Occupational Research Agenda (NORA) in 1996 to promote swift transfer of research into occupational environments to benefit the worker (NIOSH, 2011). NORA is a public–private partnership to establish priorities for occupational safety and health research at NIOSH and the country. NORA focuses on priority-driven research topics and in its first decade advanced knowledge in key scientific areas. A sector-based approach is used to allow focus on the different needs of each industry area. The NORA sector groups are listed in Table 1.2.

After an initial 10 successful years, NORA established sector councils and roadmaps. The focus is now shifting to the national research agenda. The national

TABLE 1.2
NORA Sector Group

Agriculture, forestry, and fishing
Construction
Health-care and social assistance
Manufacturing
Mining
Services
Transportation, warehousing, and utilities
Wholesale and retail trade

Source: http://www.cdc.gov/niosh/NORA/SECTOR.HTML

research agenda will focus on problems in the major industrial sectors of agriculture, construction, health care, manufacturing, mining, services, trade, and transportation. NIOSH facilitates the work of NORA to ensure that the research activities are relevant to the problems of today's workplaces. Additionally, NIOSH serves as a steward of NORA to ensure that the research is conducted using the highest quality science and that the outcomes have a measurable impact on improving the lives of workers.

1.4.12.4 Mine Safety and Health Administration

The Mine Safety and Health Administration (MSHA) is an agency of the U.S. Department of Labor with the responsibility of administering the provisions of the Federal Mine Safety and Health Act of 1977 as amended by the Mine Improvement and New Emergency Response (MINER) Act of 2006. The mission of the agency is

> to enforce compliance with mandatory safety and health standards as a means to eliminate fatal accidents; to reduce the frequency and severity of nonfatal accidents; to minimize health hazards; and to promote improved safety and health conditions in the Nation's mines.

Mines can be hazardous environments and in addition to the inherent risks of the task, the potential for flood, explosion, and collapse have the potential to impact miners and many others in this industry. The MSHA has actively sought to improve the safety of miners, particularly after recent high-profile mining fatalities in the United States and globally. This has included the enactment of legislation, enforcing stronger safety standards and requiring improved equipment, as well as the development of emergency response plans by mining companies.

1.4.12.5 Nuclear Regulatory Commission

The U.S. Nuclear Regulatory Commission (NRC) is an independent agency and was established by the Energy Reorganization Act of 1974. Congress' objective in creating the NRC was to enable the nation to safely use radioactive materials for beneficial civilian purposes while ensuring that people and the environment are protected. The mission of the NRC is as follows (NRC website, 2010):

> To regulate the nation's civilian use of byproduct, source, and special nuclear materials to ensure adequate protection of public health and safety, to promote the common defense and security, and to protect the environment.

This includes employees' safety in the nuclear industry, which involves the use of ergonomic and safety principles to promote a healthy work environment.

On March 28, 1979, the debate over nuclear power safety moved from the hypothetical to reality. An accident at Unit 2 of the Three Mile Island plant in Pennsylvania resulted in the melting of about half of the reactor's core. This incident generated fear that widespread radioactive contamination would result. The accident was attributed to many issues including a need for improved human factors principles in system design and task performance.

Fortunately, this crisis ended without a major release of dangerous forms of radiation and there was not a need to order a general evacuation of the area. Additionally, there were no deaths or immediate injuries to plant workers or members of the nearby community. Long-term health consequences are still being studied. The outcome of the Three Mile Island incident was sweeping changes involving emergency response planning, reactor operator training, human factors engineering, radiation protection, and many other areas of nuclear power plant operations. This substantially increased the NRC's recognition of the need for stronger application of human factors and ergonomic principles.

1.4.12.6 Department of Defense Military Standard (MIL-STD) 1472, 1999 (F)

The series of MIL-STDs 1472, promulgated in 1989, refers to the standards in human factors and ergonomic-related design and usability issues for the U.S. Department of Defense. The Department of Defense Design Criteria Standard is approved as the basis for general human engineering design criteria for military systems, subsystems, equipment, and facilities (MIL-STD 1472F, 2009).

There are numerous components to this standard as well as many related military standards. According to the DoD, the purpose of this standard is

> … to present human engineering design criteria, principles, and practices to achieve mission success through integration of the human into the system, subsystem, equipment, and facility, and achieve effectiveness, simplicity, efficiency, reliability, and safety of system operation, training, and maintenance.

Although it is designed for military application, MIL-STD-1472F is considered by some to be the preeminent human engineering document in the world, because it is so frequently referenced by government agencies and contractors both domestically and internationally. In fact, this standard was the basis for other human factors standards such as the FAA's *Human Factors Design Standard*, the Department of Energy's (DOE) *Human Factors and Ergonomics Handbook for the Design for Ease of Maintenance*, and the British Defence Standard 00-25, *Human Factors for Designers of Systems* (HFE Tag, 2009). MIL-STD-1472 has been acknowledged worldwide as an authoritative source for human factors requirements and design criteria.

1.4.12.7 ISO9000 Series

The International Organization for Standardization ISO is recognized globally as a leader in developing and determining quality systems for various industries including the automotive, defense, and health-care industries (ISO website, 2011). The official U.S. representative to ISO is the American National Standards Institute (ANSI). ISO has developed a series of guidelines in specific areas with a common goal of establishing and promoting standards that enhance ergonomics, safety, quality, and technical compatibility (i.e., ISO 9241 series). ISO has also published standards in human factors and ergonomics that focus on specific industries (i.e., ISO/IEC 26513:2009). These guidelines are particularly useful in establishing common practices for ergonomic applications in projects that have international participants in the development of a product, process, or environment. For example, many large corporations manufacture products (automobiles, software, etc.) in different countries, and the use of ISO standards corporate-wide promotes the standardization of processes, task activities, and software development to ultimately improve efficiency, product quality and safety practices.

1.4.13 HUMAN-CENTERED AND HUMAN-INTEGRATED DESIGN

The application of ergonomics in design is user centered design. The core of user-centered design (UCD) is designing systems, products, and processes that are focused on the user. The implementation of a UCD process requires the active participation of real users, as well as an iteration of design solutions. UCB has been used extensively in the design of websites (http://www.usability.gov/guidelines/). Numerous government agencies (Usability.gov, 2011) have applied these principles to web pages, websites, and other electronic media (NASA Usability Toolkit, 2011).

The foundation for the application of UCD on an international basis is ISO 13407: Human-Centered Design Process (ISO/TR, 16982:2002, 2002). This standard defines a general process for integrating human-centered processes throughout the development life cycle. The method to apply UCD in product design is not specified by ISO 13407, but the required activities that are to be a part of the process are outlined.

The approaches to UCD vary across industries and geographic regions. Three broad categories that can be used to classify these approaches include cooperative design, participatory design, and contextual design. A definition of these approaches is offered in the following text:

* *Cooperative Design*—Involves designers and users equally. This approach is in the Scandinavian tradition of Information Technology design, and it has been evolving since 1970 (Greenbaum and Kyng, 1991).
* *Participatory Design (PD)*—A North American term for the same concept, inspired by Cooperative Design, focusing on the participation of users (Schuler and Namioka, 1993). It is an approach to design that attempts to actively involve the end users in the design *process* to help ensure that the product designed meets their needs and is *usable*.
* *Contextual Design*—"customer-centered design," including some ideas from Participatory Design (Beyer and Holzblatt, 1996). *Context* is the circumstances under which a device is being used, as in the current occupation of the user.

Each of these approaches follows the ISO 13407 Model. The gathering of user input requires engagement and interaction with the representatives of the intended user population. A number of approaches have been used to collect this information from users including focus groups, usability testing, observation analysis, questionnaires, and interviews.

To meet the requirements of the standard, a method for UCD must include four essential elements:

- *Requirements and planning*—Understanding and specifying the context of use for the product
  - Identify intended users.
  - Identify what the product will be used to accomplish.
  - Identify the conditions under which the product will be used.
- *Requirements specification*—Specifying the user and organizational requirements
  - Define user goals that must be met.
  - Define business requirements.
- *Design*—Producing designs and prototypes
  - Develop design in stages and ideally allow users to evaluate products at different stages of development.
- *Evaluation*—Carrying out user-based assessment
  - Evaluate product by means of users that are representative of the intended population.
  - Evaluation can be an iterative process to allow multiple evaluations as revisions are made to the product.

Upon execution of the four categories of UCD, the process is complete and the design is ready to be released. Successful implementation of UCD standards has been shown to increase user acceptance and product quality and enhance profitability in many environments.

1.4.14 Cost Justification for Ergonomic Program

The motivation for ergonomic research and applications should be strong given the potential of productivity increases, quality improvements, and cost savings as a result of ergonomically designed environments. However, this is not always the case as many organizations are unwilling to make the investment. The economic benefits of ergonomic programs have been extensively studied (Alexander, 1998, Douphrate, 2004; Hendrick, 1997; MacLeod, 1995, Oxenbugh, 1997). In an applied ergonomics environments, whether it is for the investment in research, resources, or equipment, an ergonomist must always be prepared to make the financial case for ergonomics (Pulat and Alexander, 1991).

Ergonomic investments are subject to the same scrutiny as any business decision and often times more as the benefits of these applications are not always obvious. In order to establish an organizational culture that values ergonomics broadly, the scientific and economic benefit must be demonstrated and communicated. Previous examples of benefits of ergonomic implementations were listed in an article in the *Industrial Safety and Health News* (ISHN) magazine in May 2000 that conclusively demonstrates the benefits of ergonomic interventions. A few of these examples are listed in the following text (ISHN, 2000):

- Nintendo Corporation estimates a saving of $1 million annually on an ergonomic investment of $400,000. Workers' compensation costs fell by 16%–20% and lifting-related injuries declined by 80%.

- Hewlett-Packard ergonomic interventions resulted in a 105% return on investment in 1 year.
- The Red Wing Shoe Company implemented their ergonomics program in 1985. Although the results took a few years to show, by 1989 their workers' compensation insurance premiums were reduced by 70%. Their lost workdays went down by 79%.
- 3M's ergonomic program nets an annual return of $250,000 from increased productivity.

As a practicing ergonomist, it is imperative to be prepared to make the case for any proposed applications. A process to prepare and make a case for ergonomics can be adapted from the general guidelines as follows (Pulat, 1992)

- Clearly define the issues or problems that the proposed ergonomic intervention will address. Examples of this include
  - Employee injuries.
  - Workers' compensation costs for a work area.
  - Employee turnover.
  - Poor quality.
  - Less than optimal productivity.
- Calculate (or estimate) the cost of the intervention. Consider the following categories at a minimum:
  - Costs for ergonomic equipment or software.
  - Labor to install or implement intervention.
  - Time that workers will spend learning the system and/or training.
- Use a validated ergonomic cost benefit analysis tool:
  - Examples of cost benefit analysis tools are available on the OSHA website, society websites, and some state occupational safety sites, such as the Washington State Ergonomics Cost Benefit Analysis tool developed by Rick Goggins (Washington State Department of Labor and Industries, 2009).
- Develop a preliminary implementation/transition plan to launch the ergonomic improvement.
- Schedule a phased implementation, during nonpeak times to limit interruption to the workplace and the worker.
- Determine specific individuals needed for implementation.
- Develop a training plan including written, user-friendly procedures.
- Anticipate and plan for any new challenges or issues that the intervention will create.
- Analyze and describe the financial implications of the proposed ergonomic intervention in key categories in order to provide a comprehensive cost justification. Key categories to consider include
  - Increased productivity.
  - Reduction in workers compensation costs.
  - Reduction in lost days of work.
  - Improvement in quality.

- Note the intangibles by describing the following:
  - Compatibility of ergonomic intervention with organizational goals (i.e., corporate strategic goals).
  - Impact on employee morale and job satisfaction.
  - Benefit to organizational image.
  - Improved safety.
- Present the case to decision makers using a professionally prepared presentation with supporting documentation and outside resources, if necessary.

These steps can be used as a guide to make a case for application of ergonomics-based mitigation strategies and interventions in the occupational environment. Additionally, ergonomic justification worksheets have been developed by Alan Hedge of Cornell University (Hedge, 2001). These worksheets utilize specific data such as cost of interventions, required compliance, and costs of ergonomic-related injuries. This is a useful tool in quantitatively justifying an ergonomic program.

1.4.15 ROLE OF THE ERGONOMIST

Ergonomists use information about people to make the workplace safer, more comfortable, and more productive. An ergonomist also studies tasks and how they are performed by various workers. With this information, the ergonomist, working with designers and engineers, ensures that a product or service can be used comfortably, efficiently, and safely. The product design must appeal to a broad segment of the population that will use the product—including perhaps, children, the elderly, and the disabled. An ergonomist can also assess existing products and services showing where they fail to "fit" the user and suggesting how this fit may be improved.

Summary of core competencies and activities for ergonomist (IEA, 2009) are a follows:

- Investigates and analyzes the demands for ergonomics design to ensure appropriate interaction between work, product, and environment, as the factors compare to human needs, capabilities, and limitations
- Analyzes and interprets findings of ergonomics investigations
- Documents ergonomics findings appropriately
- Determines the compatibility of human capabilities with planned or existing demands
- Develops a plan for ergonomics design or intervention
- Makes appropriate recommendations for ergonomics changes
- Implements recommendations to improve human performance
- Evaluates outcome of implementing ergonomics recommendations
- Demonstrates ethical and professional behavior

These steps are a general guideline and can be adapted to meet the specific needs for an ergonomist.

1.5 SUMMARY

As can be seen from IEA's definition, ergonomics also seeks to optimize overall performance. Ergonomics fits the task to the worker, not just to prevent injury, but also to increase productivity. Safety is intertwined with physical ergonomics topics such as defining optimal work postures and workstation layouts, minimizing repetitive motions, preventing musculoskeletal disorders, and facilitating manual material handling. However, ergonomics also includes cognitive and organizational topics, which expand beyond the limited bounds of safety, to include topics such as easing mental workload, improving human–computer interaction, optimizing policies and procedures, and ensuring quality.

This chapter has provided an introduction to the field of ergonomics including the history, foundation of the discipline, and principles for application. The remaining sections of the text will expand on much of the information and prepare the student to become a researcher, practitioner, or provider of ergonomic solutions.

Case Study

Lessons from the BP Oil Spill

By Peter Budnick
June 15, 2010

People all over the world watch helplessly as the oil spill in the Gulf of Mexico goes from bad to worse. Since the initial explosion and fire, which killed 11 people and injured 17 others before the first drop of oil hit the sea, I've been left with a "we told you so" feeling.

You see, BP has a well documented history of risk management problems. Here's a sampling of articles we've published, spawned by BP's 2005 Texas City refinery explosion that killed 15 people and injured 170.

Appearing in *Ergonomics Today*™:

- **Factors Cited in Texas Explosion Point to Ergonomic Issues**, August 29, 2005
  In an unprecedented move, the US Chemical Safety and Hazard Investigation Board (CSB) issued an urgent recommendation on August 17 that requests BP create an independent panel to review the safety culture at the company's North American operations.
- **Experts Not as Quick as BP to Blame Workers for March 23 Refinery Explosion**, November 11, 2005
  Experts outside the company see other causes, some that add up to a failure to observe fundamental ergonomics principles.

- **BP North America Issues Final Report on Deadly Refinery Explosion**, December 19, 2005

   British Petroleum (BP) recently laid most of the blame for a deadly explosion at one of its five refineries in the United States on circumstances some experts could interpret as corporate failings at the macro-ergonomic level.

- **Safety and the Energy Industry—Continents Apart?** January 18, 2006

   Is Safety a Dirty Word in the Energy Industry? A Look at the Obstacles to Reform.

- **Investigations Into Deadly BP Refinery Explosion Uncover Repeated Problems**, November 9, 2005

   The Ergonomics Report™ reviewed several of the interim investigations in November and talked to experts to see if the blame is misplaced.

- **Inquiry into BP Refinery Explosion Blames Corporate Culture**, January 22, 2007

   The report of the Baker Panel, investigating the explosion at a BP refinery in Texas in 2005, was released in January. It points the finger where it doesn't often point—at the executive suite.

- **Robust Probe Finds British Petroleum Safety Culture Lacking**, February 11, 2007

   The decision makers at British Petroleum forgot about the inherent dangers of processing crude oil. The panel led by former US Secretary of State James Baker said this and more about the safety culture and processes at the company in a strikingly robust report released in January.

These articles point to a systematic breakdown in corporate decision making, risk management, and corporate culture. Professional ergonomists, who are trained to take a systems viewpoint, recognize these as macroergonomic concerns.

These kinds of accidents are predictable, and therefore preventable. Had BP taken effective steps to improve, this most recent disaster likely would not have occurred, or at least the outcome wouldn't have been so severe.

"We told you so!" Ah, but little good comes from saying "we told you so." It won't fix what's broken, and it only raises the ire of those at blame, puts them in a defensive posture, and does little to motivate them to get on with the important work of dealing with the root causes and remediating the damage.

I was reminded of this when I read *Workplace Safety is the Leading Edge of a Culture of Accountability*, by David Maxfield, writing in EHSToday.com. The article focuses on accountability, presenting a case that:

> ... once accountability for safety is reached, companies can leverage that learning to improve quality, production, cost control and customer service.

He goes on to explain the nuts and bolts of effective accountability, and how companies might build a culture of accountability.

ACCOUNTABILITY IS NOT ABOUT BLAME
Some leaders believe accountability is all about blame and punishment. Find
the guilty party and punish him or her.

Blame and punishment will always occur, but that's not the makings for an effective culture of accountability. There's already a frenzy of politicians, pundits, lawyers and corporate apologists pointing fingers of blame, and that's sure to continue for a long time to come. Unfortunately, none of that will fix BP's internal problems, nor the underlying root causes of the accident.

Maxfield starts his article with an example that's familiar to ergonomists:

"What do you want me to do, save money or save lives? You can't have it both ways."
This quote comes from a frustrated manager who feels whipsawed by these competing
values. Of course he knows the company line, "Safety comes first," but then he adds,
"But we're not in business to be safe. We're in business to build product."

Actually, you can have it both ways, and you'll find that you'll be more successful at building product—or extracting, refining and selling petroleum based products in BP's case—if you follow these philosophically simple, albeit operationally challenging steps:

1. Management Leadership
2. Respect for People (employees, customers, suppliers, stakeholders, communities)
3. A Scientific Approach to Continuous Improvement (e.g., Plan-Do-Check-Act, Standardized Methods)
4. Cross-Functional Cooperation and Teamwork
5. Training, Knowledge and Skill Development
6. Responsibility and Accountability (throughout the organization)

These characteristics are the makings of any effective, sustainable world-class organization. We could add a few things to the list, but these are the basic essential ingredients. If one or more of these features goes missing, the organization will certainly underperform, or worse, find itself in a position like BP.

Readers familiar with the Toyota Production System, which influenced what is more generally known as Lean Manufacturing (or Lean Enterprise, or Lean Journey), will recognize much of this terminology. Only those companies that have truly embraced Lean, and are enjoying the fruits of that journey, will recognize Respect for People as the key feature for success. Without it, Lean is merely a set of methods and "flavors of the month"; with it, Lean can become a systematic formula for success.

I've worked with companies on the Lean Journey and presented at various conferences to deliver this message: Ergonomics *is* Respect for People, it's essential for any company wishing to improve, and it's the secret sauce that makes it possible to have it your way—better quality, higher productivity, satisfied customers, stronger financial performance, *and* safer and more satisfying work (apologies to the fast food industry for plagiarizing their ads).

I'm not talking about ergonomics as solely an MSD prevention strategy, I'm talking about ergonomics in its broad sense (organizational, physical and cognitive ergonomics). But even a successful MSD ergonomics process/program incorporates these same essential ingredients, and provides a blueprint for, and a shining example of, a world class process.

There are two extreme approaches that companies may choose:

1. Forge ahead with little respect for people and waste enormous amounts of cash and resources to cover-up their mistakes and the messes they willingly create.
2. Or do it right the first time, minimize errors and mistakes, take responsibility for any mistakes or messes they do create, and understand and address root causes so future mistakes don't occur.

Start with ergonomics—Continuous Improvement and Respect for People—It's really that simple.

BP can't go back and undo the damage it's already done, but if they survive this debacle, they can take these lessons forward. The many other companies that admit and accept that they are accidents waiting to happen should also take heed.

Macroergonomic problems lead to far worse outcomes than a poor safety record and could spell the death of an entire company, not just the employees, shareholders, suppliers and citizens—and even ecosystems—who happen to be direct victims in the latest system failure.

Source: Total contents of Case Study reprinted with permission from Ergoweb. com, 2010.

EXERCISES

1.1 Describe the origin of the field of ergonomics.
1.2 Join an Internet-based ergonomics discussion group and seek to generate debate on a controversial ergonomic area of interest to you.
1.3 Search the web for sources of information about ergonomics. Identify sources from at least three countries and explain the commonalities and differences in the definition, application, and perception of the importance of ergonomics.
1.4 Identify three different manufacturers across a range of industries. Find out from them (or via website and other sources) whether they use ergonomic principles in the design of their products. For those companies that do use ergonomics, find out who is responsible and how ergonomics is applied.
1.5 Contact three different service industries (i.e., restaurant, grocery store). Ask if they use ergonomics in the design of the tasks, equipment selection, and processes.
1.6 Describe the present scope and concerns of the field of ergonomics. Obtain back issues for the last 5 years of two or more of the journals *Human Factors*,

Ergonomics, and *International Journal of Industrial Ergonomics and Applied Ergonomics*. Discuss any changes in the discussion of areas of need in ergonomics based on growth in technology of products, work environments, and recreational environments.

1.7 Based on the review of legislation, industry practices, and incidents, explain the extent to which you believe the following industries are aware of the need for ergonomics in their short-term or long-term planning. What mechanisms do they have in place to ensure that ergonomics principles are applied?

- Meat packing industry
- Mining
- International petroleum (oil and gas) industry
- NASA
- Health care

REFERENCES

American Psychological Association, Public Policy Office. (2009). Ergonomics: The science for better living and working. APA Online. http://www.apa.org/ppo/issues/sergofact. html (accessed August 29, 2009).

Beyer, H. and Holtzblatt, K. (1996). *Contextual Design: Defining Customer-Centered Systems*, Morgan Kaufmann: San Francisco, CA.

Diseases of Workers. (2009). Encyclopedia Britannica Online. http://www.britannica.com/ EBchecked/topic/165553/Diseases-of-Workers (accessed August 28, 2009).

ErgoWeb. (2010). The role of the ergonomist. http://www.ergoweb.com/news/news_list. cfm?cat_id=1 (accessed July, 2011).

Greenbaum, J. and Kyng, M. (1991). *Design at Work: Cooperative Design of Computer Systems*, Lawrence Erlbaum: Hillsdale, NJ.

Helander, M. (2005). *A Guide to Human Factors and Ergonomics*, Vol. 59, Taylor & Francis: New York.

International Ergonomics Association (IEA). (2009). http://www.iea.cc/browse. php?contID=international_ergonomics_association (accessed August 29, 2009).

ISO Website. (2011). Ergonomics of human-system interaction—Usability methods supporting human-centred design. http://www.iso.org/iso/search.htm?qt=ergonomics&publish ed=on&published=on&active_tab=standards

Koppes, L.L. (Ed.) (2006). *Historical Perspectives in Industrial and Organizational Psychology*, Lawrence Erlbaum Associates: Mahwah, NJ, p. 245.

Making a business case: Selling ergonomics. *Industrial Safety & Hygiene News Magazine*, May 20, 2000.

NASA AMES. (2011). Human Systems Integration Division. http://human-factors.arc.nasa. gov/ (accessed March 2011).

NASA's Usability Toolkit, 2011. http://www.hq.nasa.gov/pao/portal/usability/overiew/index.htm

OSHA Website. (2011). (accessed September 5, 2011).

http://www.cdc.gov/niosh/about.html (accessed February 13, 2010).

http://www.ergoweb.com/resources/casestudies/ (accessed July 2010).

Price, B. (1989). Frank and Lillian Gilbreth and the manufacture and marketing of motion study, 1908–1924. Business and Economic History, Second Series, 18. *The Business History Conference*, Wilmington, DE, March 31–April 2.

Pulat, B.M. (1992). *Fundamentals of Industrial Ergonomics*, Prentice Hall: Englewood Cliffs, NJ.

Pulat, B.M. and Alexander, D.C. (Eds.) (1991). *Industrial Ergonomics: Case Studies*, Institute of Industrial Engineers: Norcross, GA.

Salvendy, G. (ed.) (2006). *Handbook of Human Factors and Ergonomics*, John Wiley & Sons: Hoboken, NJ.

Schuler, D. and Namioka, A. (1993). *Participatory Design*, Lawrence Erlbaum: Hillsdale, NJ, and chapter 11 in Helander et al.'s Handbook of HCI, Elsevier 1997.

Stramler, J.H. (1993). *The Dictionary for Human Factors/Ergonomics*, CRC Press: Boca Raton, FL.

Taylor, F.W. (1911). *The Principles of Scientific Management*, Harper Bros: New York.

Usability Professionals Association. http://www.upassoc.org/usability_resources/about_usability/what_is_ucd.html (accessed March 11, 2011).

Washington State Department of Labor and Industries. (2009). Cost benefit analysis—Instructions for use. http://www.pshfes.org/cba.htm (accessed August 29, 2009).

Webcredible. (2010). http://www.webcredible.co.uk/user-friendly-resources/web-usability/user-centered-design.shtml (accessed June 14, 2011).

Winkel, J. and Westgaard, R. (1996). Achieving ergonomics impact through management intervention. *Appl. Ergon.* 27(2), 111–117. doi:10.1016/0003-6870(95)00065-8.

Wood, M.C. and Wood, J.C. (Eds.) (2003). *Frank and Lillian Gilbreth: Critical Evaluations in Business and Management*, Routledge: New York.

Word IQ. (2010). http://www.wordiq.com/definition/Contextual%20design (accessed May 9, 2011).

LISTING OF ERGONOMIC RESOURCES

- Human Factors and Ergonomic Society: www.HFES.org
- The International Ergonomics Association: www.IEA.cc
- Occupational Safety and Hazard Administration/Ergonomics: www.osha.gov/SLTC/ergonomics/index.html
- Cornell University Ergonomics: ergo.human.cornell.edu/
- Centers for Disease Control: www.cdc.gov/niosh/
- CTD News: www.ctdnews.com
- Ergonomic Information Analysis Center—University of Birmingham: www.eee.bham.ac.uk/eiac/index.htm
- Liberty Mutual Research Institute for Safety: www.libertymutualgroup.com
- National Council on Compensation Insurance (NCCI): www.ncci.com
- OSH-DB Occupational Safety and Health Database, University of Wisconsin, Whitewater: library.uww.edu/subject/safety.html
- University of Michigan Center for Ergonomics: www.engin.umich.edu/dept/ioe/C4E

ERGONOMICS AND RELATED ASSOCIATIONS

ACGIH, American Conference of Government Industrial Hygienists
6500 Glennway Avenue
Cincinnati, OH 45221, USA

ANSI, American National Standards Institute
11 West 42nd Street, 13th Floor
New York, NY 10036, USA

ASHRAE, American Society of Heating, Refrigerated, and Air-conditioning Engineers
1791 Tullie Circle
Atlanta, GA 30329, USA

CSERIAC, Crew System Ergonomics Information Analysis Center
AL/CFH/CSERIAC
Wright-Patterson AFB, OH 45433-6573, USA

Ergonomics Society
Devonshire House
Devonshire Square
Loughborough, Leics., 11 3DW, UK

Human Factors and Ergonomics Society
P.O. Box 1369
Santa Monica, CA 90406, USA

Military (USA)
U.S. Military and Federal Standards, Handbooks, and Specifications are available from:

National Technical Information Service, NTIS
5285 Port Royal Road
Springfield, VA 22161, USA

Or

Naval Publications and Forms Center, NPODS
5801 Tabor Avenue
Philadelphia, PA 19120-5099, USA

Or

Standardization Division, U.S. Army Missile Command, DRSMI-RSDS
Redstone Arsenal, AL 35898, USA

Or

U.S. Air Force Aeronautical Systems Division, Standards Branch
ASD-ENESS
Wright-Patterson AFB, OH 45433, USA

NASA, National Aeronautics and Space Administration SP 34-MSIS
LBJ Space Center
Houston, TX 77058, USA

National Safety Council
444 North Michigan Avenue
Chicago, IL 60611, USA

NIOSH, National Institute for Occupational Safety and Health
4676 Columbia Parkway
Cincinnati, OH 45226, USA

SAE, Society of Automotive Engineers
400 Commonwealth Drive
Warrendale, PA 15096-0001, USA

ISO, International Organization for Standardization
1 rue Varembe
Case Postale 56
CH 1211 Geneve 20, Switzerland

ADDITIONAL SOCIETIES AND PROFESSIONAL ORGANIZATIONS

American Industrial Hygiene Association (AIHA)
American Society of Safety Engineers (ASSE)
Association of Canadian Ergonomists
ASTM International (ASTM)
Board of Certification in Professional Ergonomics (BCPE)
Chemical Safety and Hazard Investigation Board (CSB)
Ergonomics Society (London)
Ergonomics Society of Australia
Human Factors and Ergonomics Society
Indian Society of Ergonomics
Institute of Industrial Engineers
International Ergonomics Association
International Society for Occupational Ergonomics
Safety Institution of Occupational Safety and Health (United Kingdom)
International Society of Biomechanics
Joint Commission of Accreditation of Healthcare Organizations
National Fire Protection Association
TUV Rheinland of North America

GOVERNMENT ORGANIZATIONS

Bureau of Labor Statistics—Health and Safety Data
Canadian Centre for Occupational Health and Safety
U.S. Department of Energy
U.S. Department of Labor
U.S. Department of Transportation
Environmental Protection Agency
U.S. Department of Health and Human Services
Mine Safety and Health Administration
National Institute for Occupational Safety and Health
National Personal Protective Technology Laboratory
Nuclear Regulatory Commission

National Transportation Safety Board
Occupational Safety and Health Administration (OSHA)

STANDARDS

American National Standards Institute (ANSI)
Canadian Standards Association (CSA)
European Committee for Standardization
International Labor Organization
International Organization for Standardization
National Institute of Standards and Technology
Standards Council of Canada

2 Systems of the Human Body

2.1 LEARNING GOALS

This chapter will provide the student with knowledge of the systems of the human body. A basic explanation and introduction is offered to establish a foundation for understanding and applying ergonomic principles in a manner consistent with the needs of the human.

2.2 KEY TOPICS

- Definitions for systems of the body
- Purpose of each system

2.3 INTRODUCTION AND BACKGROUND

The scientific study of the human body began as early as the fifth century BCE in Greece, the center of academia at the time. According to records, the first researcher to dissect a human body for the purpose of studying it was Alcmaeon of Crotona, who was searching for the organic factor responsible for human intelligence (Huffman, 2008). While, at the time, it was widely believed that the heart was responsible for thought and awareness, Alcmaeon was the first to propose that the intellect resided in the head. He observed that there seemed to be a loss of reasoning capabilities in those who have suffered head trauma, and concluded that this must be the source of reason. By the third century BCE., two famous surgeons, Herophilus and Erasistratus, continued the use of dissection to expand the field of anatomy. Specifically, they were granted permission to perform *vivisections* on criminals, whose sufferings were considered ethical and justified in the name of both science and religion.

By the turn of the sixteenth century, the artistic community contributed to Europe's growing knowledge of the human body. Around 1489, Leonardo da Vinci began his well-known series of exceptionally accurate anatomical drawings, culminating in 750 drawings from the dissection of an estimated 30 corpses. In 1628, Harvey Williams published *Exercitatio anatomica de motu cordis et sanguinis in animalibus* (*The Anatomical Function of the Movement of the Heart and the Blood in Animals*), which argued against many anatomical misconceptions of the day, but most importantly, that blood does not drift about the human body at random, but actually circulates in a very specific rhythm. This book became the standard for anatomical literature until Drs. Henry Gray and Henry Vandyke Carter published *Anatomy: Descriptive and Surgical*, informally known as *Gray's Anatomy*, in 1858 (Gascoigne, 2001).

2.4   SYSTEMS OF THE BODY

The human body is a complex organism with ten essential body systems, including the circulatory, respiratory, skeletal, muscular, nervous, immune, reproductive, urinary, digestive, and integumentary systems. The description and images of the systems are based on those proved by Gray (1918). A body system is defined as a group of organs working together to provide a specific function. While all systems are clearly important to overall function of the body, only six will be focused on for the purposes of this ergonomics text. The human body systems emphasized in this chapter are as follows:

- Circulatory and cardiovascular systems (subsystem of the circulatory system)
- Respiratory system
- Skeletal system
- Muscular system
- Nervous system
- Integumentary (skin) system

The systems perform an array of singular and overlapping functions that allow the human body to operate. The following sections provide a brief overview of each of these systems. Many figures and discussions of these systems are from ADAM (2011).

2.4.1   CIRCULATORY SYSTEM

The circulatory system consists of the organs and tissues involved in circulating blood and lymph through the body. The cardiovascular system can be considered a subsystem of the circulatory system. The cardiovascular system in its simplest form is a closed loop system, meaning that the blood never leaves the network of the blood vessels; this system begins and ends at the heart. The blood is in a continuous flow as it circulates through the loop repeatedly. The cardiovascular system transports vital materials throughout the body carrying nutrients, water, and oxygen to the cells and carrying away waste products such as carbon dioxide.

The cardiovascular system comprises five major components: the heart, arteries, arterioles, capillaries, and veins. Each of these elements of the circulatory system is discussed in more detail in the following text.

2.4.1.1   Heart

The heart is the most recognized component of the cardiovascular system and comprises thick muscle tissue. It is located in the center of the chest cavity slightly to the left of the sternum and is about the size of a fist. The heart beats approximately three billion times during an average person's lifetime. The heart is the main pump of the cardiovascular system. It is divided into four chambers, the top two chambers are known as atria, and the bottom two chambers are called ventricles. The atria both contract at the same time as do the two ventricles. Blood enters the heart via the superior and the inferior vena cava. These are the two largest veins in the body. The right atrium receives the blood first. The right atrium contracts and forces the blood

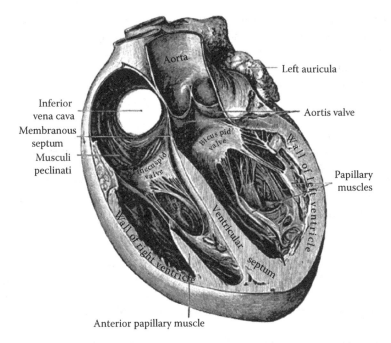

Aorta

Left auricula

Inferior
vena cava

Aortis valve

Membranous
septum

Bicuspid
valve

Musculi
peclinati

Tricuspid
valve

Wall of left ventricle

Papillary
muscles

Wall of right ventricle

Ventricular

septum

Anterior papillary muscle

FIGURE 2.1 Heart. (From Gray, H., *Anatomy of the Human Body*, Lea & Febiger, Philadelphia, PA, 1918, on-line edition published May 2000 by Bartleby.com; © 2000 Copyright Bartleby.com, Inc.)

into the right ventricle. Then the blood is pumped out of the heart to the lungs. The complete process is described following the explanation of the components of the circulatory system (Figure 2.1).

2.4.1.2 Arteries

The arterial system is the higher-pressure portion of the cardiovascular system. Arterial pressure varies between the peak pressure during heart contraction, called the systolic pressure, and the minimum or diastolic pressure between contractions, when the heart rests between cycles. These pressures determine the commonly recognized term "blood pressure" that represents the intensity of the blood flowing through the heart and other areas of the circulatory system. The arterial system consists of a complex collection of arteries and arterioles that carry blood away from the heart. All arteries, with the exception of the pulmonary and umbilical arteries, carry oxygenated blood throughout the body. The arterial system is divided into systemic arteries, carrying blood from the heart to the whole body, and pulmonary arteries, carrying blood from the heart to the lungs.

2.4.1.3 Arterioles

An arteriole is a small diameter blood vessel that extends and branches out from an artery and leads to capillaries. Arterioles are smaller than the arteries and regulate the flow of blood in various parts of the body. These structures support body functions such as digestion, thermoregulation, and many other critical functions. For example, in order to reduce internal heat loss when the body is cold, the arterioles leading to the skin will narrow to lessen the flow of blood to the skin. Similarly, the arterioles leading to the digestive system expand during digestion to increase the energy provided to that system.

2.4.1.4 Capillaries

Connecting the arterioles and venules (small blood vessels that transport deoxygenated blood from the capillary beds to veins), the capillaries are the smallest of the body's blood vessels measuring approximately 5–10 μm in diameter. The very small vessels enable the interchange of water, oxygen, carbon dioxide, nutrients and waste. The capillaries also enable the exchange of chemical substances between the blood and the surrounding tissue.

2.4.1.5 Veins

As previously described, the function of the arteries is to transport oxygenated blood to the muscles and organs of the body, where its nutrients and gases are exchanged at capillaries. The veins complete this very important cycle within the circulatory by transporting the deoxygenated blood from the organs and tissues back to the heart (Figure 2.2).

The following steps summarize the flow of blood through the cardiovascular system (Figure 2.3):

- As blood moves through the heart, it passes through each of the four chambers (upper right, lower right, upper left, and lower left), takes a detour to the lungs to remove waste (carbon dioxide) and oxygenate, and then proceeds to the lower left-hand chamber called the left ventricle.
- When the right ventricle contracts, the blood is pumped into both lungs via the pulmonary artery. This portion of the circulatory system is sometimes referred to as pulmonary circulation or lesser circulation. The pulmonary artery is the only artery in the body that carries deoxygenated blood.
- Blood is returned from the lungs via the pulmonary veins. These are the only veins in the body that carry oxygenated blood. The oxygenated blood is returned to the left atrium. Blood is sent into the left ventricle when the atrium contracts. The left ventricle is the strongest and most muscular portion of a healthy heart.
- The left ventricle pumps forcefully to circulate blood throughout the entire body. Blood is forced into the aorta, the main artery leaving the heart, when the left ventricle contracts.
- The oxygenated blood is now forced throughout the body through a series of arteries that gradually become smaller and smaller. Blood flows from the arteries into arterioles and then from arterioles into capillaries. At this point, the blood is delivered to the cells via the blood's contact with the interstitial intercellular fluid, and waste products are collected.

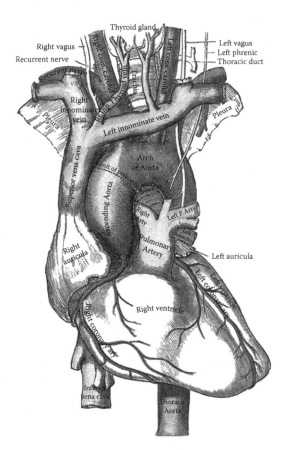

FIGURE 2.2 How it functions. (From Gray, H., *Anatomy of the Human Body*, Lea & Febiger, Philadelphia, PA, 1918, on-line edition published May 2000 by Bartleby.com; © 2000 Copyright Bartleby.com, Inc.)

- The blood now starts its return trip to the heart. From the capillaries, the blood flows into venules. These are very small veins about the same size as the capillaries. From the venules the blood is then sent into the veins.
- The veins return the blood either into the superior or the inferior vena cava. The superior vena cava is the vein that carries deoxygenated blood from the upper half of the body to the heart's right atrium. As the name suggests, the inferior vena cava is the vein that serves the lower half of the body, by transporting deoxygenated blood from the lower half of the body into the right atrium of the heart. The liver and kidney also support the circulatory system by removing waste products from the blood.

The complex nature of the circulatory and cardiovascular systems provide the essential functions of oxygen delivery, removal of waste, and distribution of fuels to various aspects of the body. These systems interact with, and support, the other major

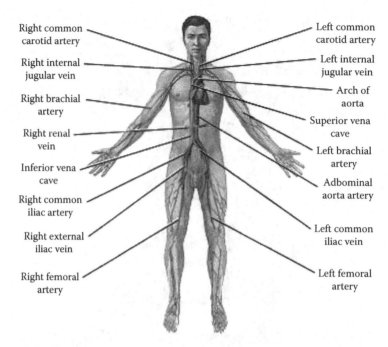

Right common carotid artery

Right internal jugular vein

Right brachial artery

Right renal vein

Inferior vena cave

Right common iliac artery

Right external iliac vein

Right femoral artery

Left common carotid artery

Left internal jugular vein

Arch of aorta

Superior vena cave

Left brachial artery

Adbominal aorta artery

Left common iliac vein

Left femoral artery

FIGURE 2.3 Circulatory system. (From Medline Plus website: A service of the National Institute of Health, http://www.nlm.nih.gov/medlineplus/ency/imagepages/8747.htm accessed November 5, 2011.)

body systems to deliver the fuel and remove waste products in order to allow the accomplishment of bodily functions.

2.4.2 RESPIRATORY SYSTEM

The primary function of the respiratory system is gaseous exchange, which is the process of oxygenation of the blood and removal of waste products like carbon dioxide. The respiratory system integrates with other systems, particularly the circulatory system, to accomplish its goal. The respiratory system includes the mouth, nose, trachea, lungs, and diaphragm (Figure 2.4).

Consider the respiratory system interacting with the external environment to receive oxygenated air at the beginning of the process. The following steps summarize the activities that this system utilizes to accomplish the respiratory goals:

- The mouth and nose allow oxygen to enter the respiratory system. This path continues through the larynx, followed by the trachea.
- The oxygen then enters the bronchi, which are two smaller tubes that branch from the trachea.
- The bronchi are then further divided into bronchial tubes, which flow into the lungs and disperse into even smaller branches that connect to the

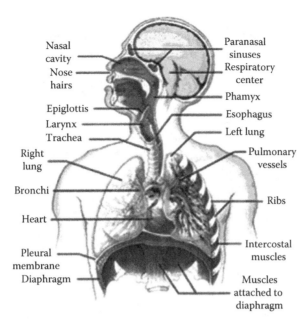

Nasal cavity
Nose hairs
Epiglottis
Larynx
Trachea
Right lung
Bronchi
Heart
Pleural membrane
Diaphragm

Paranasal sinuses
Respiratory center
Phamyx
Esophagus
Left lung
Pulmonary vessels
Ribs
Intercostal muscles
Muscles attached to diaphragm

FIGURE 2.4   How the respiratory system works. (From Phillips, 2009.)

alveoli. These are tiny, air-filled sacs that number as high as 600 million in the average adult.
- The oxygen enters the alveoli and is diffused through the surrounding capillaries. It then enters the arterial blood completing its course.

At the same time, the alveoli absorb carbon dioxide from waste-rich blood in the veins. This carbon dioxide exits the lungs upon exhaling. A simplified explanation of the events and actions in the respiratory process is represented in Box 2.1.

2.4.3 SKELETAL SYSTEM

The skeletal system consists of bones and connective tissue to create the structural foundation of the human body (Figure 2.5). The skeleton provides structural support for the body and protects the internal organs, such as the heart and spinal column, from external forces. The skeletal system can be divided into two components: the axial skeleton (spine, ribs, sacrum, sternum, cranium, and about 80 bones) and the appendicular skeleton (bones of the arms, pelvis, legs, and shoulders, totaling 126 bones). While at birth a human has about 350 bones, over time, certain bones of the body fuse together, leaving the average adult body with 206 bones. The major bones that are studied in the field of ergonomics are those of the extremities, skull, spine, and major joints. Figure 2.6 provides an illustration of some of these bones.

Bones are formed through a process called ossification. Because most of the body's calcium is contained in bones, the process of calcification is sometimes confused with ossification. Ossification is the actual activity that produces bone, while

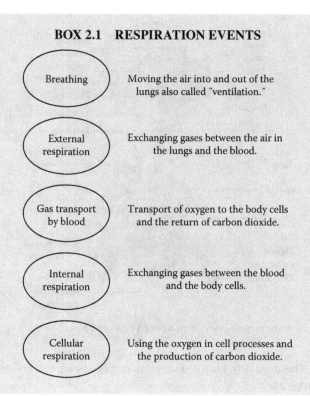

BOX 2.1   RESPIRATION EVENTS

Breathing — Moving the air into and out of the lungs also called "ventilation."

External respiration — Exchanging gases between the air in the lungs and the blood.

Gas transport by blood — Transport of oxygen to the body cells and the return of carbon dioxide.

Internal respiration — Exchanging gases between the blood and the body cells.

Cellular respiration — Using the oxygen in cell processes and the production of carbon dioxide.

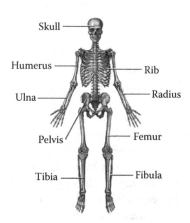

FIGURE 2.5 Human skeleton. (From Medline Plus website: A service of the National Library of Medicine, National Institute of Health, http://www.nlm.nih.gov/medlineplus/ency/imagepages/9065.htm accessed November 5, 2011.)

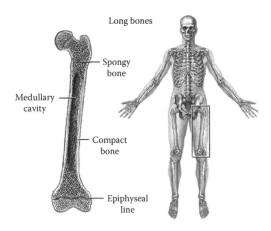

Long bones

Spongy
bone

Medullary
cavity

Compact
bone

Epiphyseal
line

FIGURE 2.6   Major bones. (From Medline Plus website: A service of the National Library of Medicine, National Institute of Health, http://www.ncbi.nim.nih.gov/pubmedhealth/ PMH0002910/figure/A002249.B9582/?report=objectonly accessed November 5, 2011.)

calcification is the formation of calcium-based salts and crystals within bones and tissues. Calcification occurs during ossification, but ossification does not occur during calcification.

Initially, the human skeleton consists primarily of cartilage but is gradually transformed into hard bones during infancy and early child development.

The body has multiple types of bones that range in size, function, and structure. The bones can be classified as long, short, axial, or irregular. Examples of each of these types of bones are provided in the following text (Figure 2.7):

1. Long bones
    a. Bones of the extremities
    b. Consists of two parts
        i. Shaft (diaphysis)
        ii. Two ends (epiphyses)
2. Short bones
    a. Smaller bones, such as carpal bones in the hand or the tarsal bones in the feet (Figure 2.8) (Gray, 1918)
3. Axial bones or flat bones
    a. Axial bones are generally flat structures, such as those in the skull, pelvis, sternum, and ribs (Figure 2.9)
4. Irregular bones (Figure 2.10)
    a. Bones such as those found in the vertebrae (Figure 2.11) (Gray, 1918)

2.4.3.1   Function of the Bones

Bone is the major structural element supporting the human body and is highly mineralized, containing approximately 99% of the body's calcium. The most important mechanical properties of bone are stiffness and strength. The primary functions of the bones are to

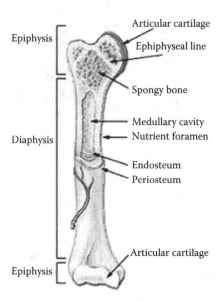

FIGURE 2.7   Long bones. (From Gray, H., *Anatomy of the Human Body*, Lea & Febiger, Philadelphia, PA, 1918, on-line edition published May 2000 by Bartleby.com; © 2000 Copyright Bartleby.com, Inc.)

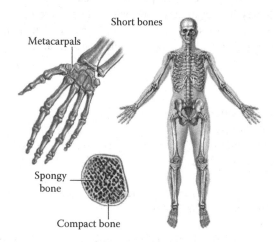

FIGURE 2.8   Short bones. (From Medline Plus website: A service of the National Library of Medicine, National Institute of Health, http://www.nlm.nih.gov/medlineplus/ency/imagepages/9889.htm accessed November 5, 2011.)

- Provide structure to the body
- Provide mechanical support and sites for muscle attachment
- Provide stability and be an instrument on which the muscular forces can be transmitted
- Protect organs housed in the cavities of the torso

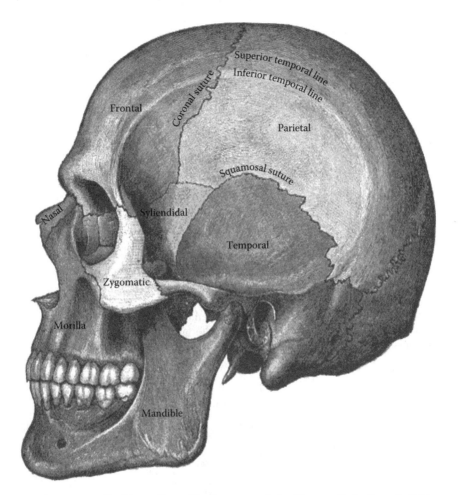

FIGURE 2.9 Skull. (From Gray, H., *Anatomy of the Human Body*, Lea & Febiger, Philadelphia, PA, 1918, on-line edition published May 2000 by Bartleby.com; © 2000 Copyright Bartleby.com, Inc.)

- Protect the central nervous system (CNS) and spine
- Protect the bone marrow
- Provide metabolic functions such as storage of calcium and potassium ions

As aging occurs (beginning around 35 years of age), the skeletal system begins to change in mineral content. There is a progressive decrease in bone mineral content and a reduction in bone thickness, making the bones more susceptible to fracture and erosion around the joints.

2.4.3.2 Bone Failure

When a bone is loaded to failure it fractures and can eventually break. Most fractures of bone occur as a result of forces that are acute in nature. However, low-level

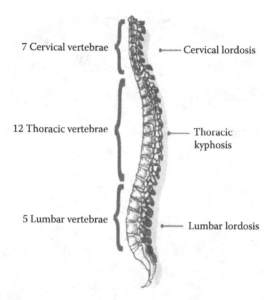

7 Cervical vertebrae — Cervical lordosis

12 Thoracic vertebrae — Thoracic kyphosis

5 Lumbar vertebrae — Lumbar lordosis

FIGURE 2.10 Irregular bones. (From Gray, H., *Anatomy of the Human Body*, Lea & Febiger, Philadelphia, PA, 1918, on-line edition published May 2000 by Bartleby.com; © 2000 Copyright Bartleby.com, Inc.)

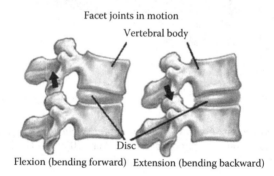

Facet joints in motion

Vertebral body

Disc

Flexion (bending forward)   Extension (bending backward)

FIGURE 2.11 Single spinal vertebrae. (From Gray, H., *Anatomy of the Human Body*, Lea & Febiger, Philadelphia, PA, 1918, on-line edition published May 2000 by Bartleby.com; © 2000 Copyright Bartleby.com, Inc.)

repetitive loading can produce fractures as well. The amount of load, level of repetition, and angle of loading are all factors that impact the likelihood of bone failure.

2.4.3.3   Vertebral Column

The vertebral column consists of 24 movable vertebrae of varying sizes that are flexible and interconnected by cartilaginous tissue. Each of these connections form a joint that allows a degree of articulation that is linked to the location and functionality of the given spine region.

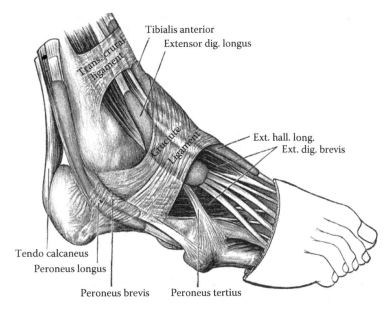

Tibialis anterior
Extensor dig. longus
Trans. crural ligament
Cruciate ligament
Ext. hall. long.
Ext. dig. brevis
Tendo calcaneus
Peroneus longus
Peroneus brevis   Peroneus tertius

FIGURE 2.12 Tendons. (From Gray, H., *Anatomy of the Human Body*, Lea & Febiger, Philadelphia, PA, 1918, on-line edition published May 2000 by Bartleby.com; © 2000 Copyright Bartleby.com, Inc.)

The regions of the vertebrae consist of the following sections (see Figure 2.12):

- Cervical (7 vertebrae)
- Thoracic (12 vertebrae)
- Lumbar (5 vertebrae)
- Sacrum and coccyx (5 fused vertebrae form the sacrum and coccyx at end of the vertebral column)

The vertebral column has four natural curves. The anterior sacral convex curve bows toward the stomach or pelvic region. The lumbar lordosis curves toward the back. The thoracic kyphosis curves toward the front of the chest, and the cervical lordosis curves toward the back of the neck (Figure 2.12 shows these curves).

The main function of the spine is to protect the spinal cord and provide vertical stabilization for the body. While the spine is a rigid structure capable of withstanding considerable load, the joints in the spine allow a range of flexibility, which is essential for human motion and posture changes. Additionally, the spine bears the mass of external loads when applied. These loads are transferred as forces and movements from the head and shoulder bones to the pelvis. Applying loads to the spinal column limits the range of mobility and flexibility.

2.4.3.4 Connective Tissue

Connective tissues, as the name implies, are components that link various structures together within the body. The most common type of connective tissue is loose connective tissue, known as the areolar tissue, which holds the organs in place. There are three other main types of connective tissue, including tendons, ligaments, and cartilage.

1. *Tendons*

   Tendons attach muscle to bone and transmit forces to produce the desired body movement. Tendons are generally surrounded by sheaths of fibrous tissue to prevent friction. The inner lining of the sheath is the synovium, also known as the synovial membrane. The synovial membrane produces synovial fluid that has a very low coefficient of friction allowing the tendons to smoothly glide through the sheath.

2. *Ligaments* (Gray, 1918)

   Ligaments connect bone-to-bone to provide the body with structural stability. The collagen contained in the ligaments results in significant tensile strength and produces the ability to withstand external forces and movements. Ligaments can be short, fat, irregular, or sheet-like structures of soft tissues (Figure 2.13).

3. *Cartilage*

   Cartilage is a soft tissue that covers articulation surfaces and is present in some organs such as the nose, respiratory tract, and intervertebral discs. There are three broad classifications of cartilage that include

   a. Hyaline cartilage
      i. Articulation surfaces
      ii. Respiratory tract
   b. Fibrous cartilage
      i. Intervertebral discs
   c. Elastic cartilage
      i. Ear
      ii. Epiglottis

The connective tissues serve various and essential functions throughout the body. These tissues can be negatively impacted by overexertion in the musculoskeletal system in the occupational environment and result in cumulative trauma disorder. These topics will be covered in the following chapters.

2.4.3.5 Joints

Areas of the body where cartilage connects two or more bones are referred to as articulations or joints. The body has multiple joints with varying degrees of articulation. Joints are classified by their *functionality*, *movement*, or *structure* (Figure 2.14).

The joints of the skull are classified as suture joints and do not have a wide range of movement, particularly as the human moves beyond childhood. These are sometimes referred to as immovable joints. These joints allow for growth but have limited flexibility. Hinge joints such as those at the knees and elbows allow movement in one plane. Ball and socket joints of the shoulder and hip can rotate in a complete circle; however, this movement may be limited by the muscular system in some instances. Finally, two flat-surfaced bones that slide over each other produce a gliding joint,

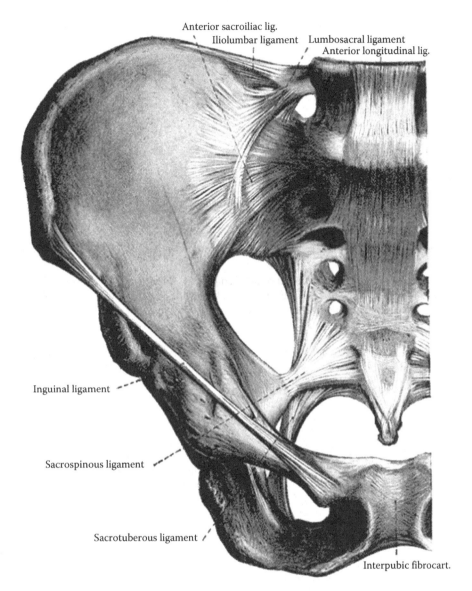

Anterior sacroiliac lig.
Iliolumbar ligament   Lumbosacral ligament
Anterior longitudinal lig.

Inguinal ligament

Sacrospinous ligament

Sacrotuberous ligament

Interpubic fibrocart.

FIGURE 2.13   Ligaments. (From Gray, H., *Anatomy of the Human Body*, Lea & Febiger, Philadelphia, PA, 1918, on-line edition published May 2000 by Bartleby.com; © 2000 Copyright Bartleby.com, Inc.)

such as those in the wrist and foot. These joints have limited movement. The two primary structural types of joints are cartilaginous and synovial. The synovial joints are the primary type of joint evaluated in ergonomic studies. Synovial joints contain a cavity with viscous (synovial) fluid. Examples include the shoulder (ball and socket joint), elbow (hinge), saddle (base of thumb), ellipsoid (wrist, ankle), pivot (neck), and gliding (i.e., intercarpal) joints (Figure 2.15).

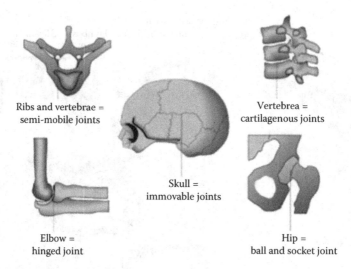

Ribs and vertebrae =
semi-mobile joints

Vertebrea =
cartilagenous joints

Skull =
immovable joints

Elbow =
hinged joint

Hip =
ball and socket joint

FIGURE 2.14   Types of joints found in the human body. (From Gray, H., *Anatomy of the Human Body*, Lea & Febiger, Philadelphia, PA, 1918, on-line edition published May 2000 by Bartleby.com; © 2000 Copyright Bartleby.com, Inc.)

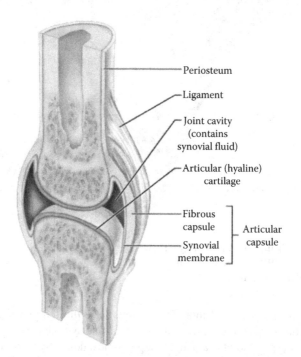

Periosteum

Ligament

Joint cavity
(contains
synovial fluid)

Articular (hyaline)
cartilage

Fibrous
capsule

Synovial
membrane

Articular
capsule

FIGURE 2.15   Synovial joint. (From Gray, H., *Anatomy of the Human Body*, Lea & Febiger, Philadelphia, PA, 1918, on-line edition published May 2000 by Bartleby.com; © 2000 Copyright Bartleby.com, Inc.)

Joint lubrication is an essential action that allows the free movement of muscles and bones. Articulating surfaces are highly lubricated through boundary or film lubrication. In boundary lubrication, the lubrication reacts with the cartilage permitting movement without adhesion. Film lubrication occurs when external pressure generates a pumping effect that sends fluid between articulating surfaces. This fluid minimizes friction and allows ease in the sliding of the joints.

2.4.4 MUSCULAR SYSTEM

The muscular system is an extensive network of muscle fibers integrated with nervous tissue and spread throughout the body. The human body contains more than 650 individual muscles with the primary function of providing movement for the body. The skeletal muscular system is controlled by the Central Nervous System, which continually sends impulses to voluntary and involuntary function areas of the muscular system (DeGraff, 1998).

Voluntary muscles are under the intentional control of the human. These muscles respond to provide support, movement, strength requirements, and other physical responses as required by physical demands. Involuntary muscles are muscles that are not directly controllable at will. These muscles are not as free as the voluntary muscles, such as those in the arms, legs, fingers, or similar regions of the body. These muscles operate automatically and function without conscious guidance from the human. Examples of involuntary muscles include the cardiac muscle and the muscles of the esophagus and intestines.

The muscular system consists of three broad categories of muscle tissue: skeletal, cardiac, and smooth. Although the heart and other voluntary muscles are impacted by ergonomic activities, the primary muscle group studied in this text is the skeletal muscle.

2.4.4.1 Skeletal Muscle

The skeletal muscle composes approximately 40%–50% of an adult's body weight. It has stripe-like markings called striations. Skeletal muscles are composed of long muscle fibers. Each of these fibers is a cell which contains several nuclei. Skeletal muscles are all voluntary muscles, because they are under control of the somatic nervous system. The main function of the skeletal muscles is to provide the power for body motion. Skeletal muscles facilitate movement, by applying force to the bones and joints through contraction.

2.4.4.2 Cardiac Muscle

The cardiac muscle exists only in the heart of the human body, where it forms the walls of the heart called the myocardium. Cardiac muscle is similar in structure to skeletal muscle, differing only in how the muscle fibers are formed and in the fact that cardiac muscles are not anchored to a bone.

2.4.4.3 Smooth Muscle

Smooth muscles are involuntary muscles under the control of the autonomic nervous system. They are responsible for the contraction of hollow organs, such as blood vessels, the gastrointestinal tract, the bladder, the uterus, and similar organs.

Its structure differs greatly from that of skeletal muscle, although it can develop isometric force per cross-sectional area that is equal to that of skeletal muscle. However, the speed of smooth muscle contraction is only a small fraction of the contractile speed of skeletal muscle. The differentiating feature of smooth muscle is the lack of visible cross striations (hence the name smooth). Smooth muscle fibers are much smaller (2–10 µm in diameter) than skeletal muscle fibers (10–100 µm). Smooth muscle is primarily under the control of the autonomic nervous system, whereas skeletal muscle is under the control of the somatic nervous system (Barany, 2002).

2.4.5 Nervous System

The nervous system provides the neural function for all aspects of the human body and is a collection of cells, tissues, and organs. The nervous system is composed of the CNS and the peripheral nervous system. The CNS includes the brain and the spinal cord. The peripheral nervous system consists of nerves that branch from the spinal cord and extend into all regions of the body (New World Encyclopedia, 2011).

The CNS is composed of billions of neurons. These neurons communicate by sending electrical signals that travel along a part of the nerve cell called an axon (nerve fiber). Electrical signals are sent along the axons at high speeds, transmitting information between nerve cells. Axons (nerve fibers) are covered with a protective fatty substance called the myelin sheath. Myelin is similar to the coating around electrical wires. This complex system continually operates to communicate between neurons to produce the desired outcomes of the nervous system.

The peripheral nervous system comprises sensory receptors, sensory neurons, and motor neurons. Sensory receptors are activated by internal or external stimuli. The stimuli are converted to electronic signals and transmitted to sensory neurons. Sensory neurons connect sensory receptors to the CNS. The CNS processes the signal and transmits a message back to an effector organ (an organ that responds to a nerve impulse from the CNS) through a motor neuron.

2.4.5.1 Central Nervous System

The CNS, consisting of the brain and spinal cord, is analogous to a central processing unit (CPU) in a computer. The CNS is responsible for controlling sensations and processing signals in different regions of the body. Functions affected by the CNS include muscle control, eyesight, breathing, memory, and others (Figure 2.16).

2.4.5.2 Peripheral Nervous System

The peripheral nervous system branches off of the spinal cord to innervate the entire body. It consists of the somatic nervous system and the autonomic nervous system. Despite their names, both systems largely execute automatically, but the autonomic nervous system is so named because it is responsible for the body's maintenance functions, which are primarily under unconscious control (Figure 2.17).

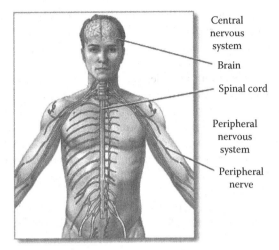

FIGURE 2.16 Central nervous system. (From Gray, H., *Anatomy of the Human Body*, Lea & Febiger, Philadelphia, PA, 1918, on-line edition published May 2000 by Bartleby.com; © 2000 Copyright Bartleby.com, Inc.)

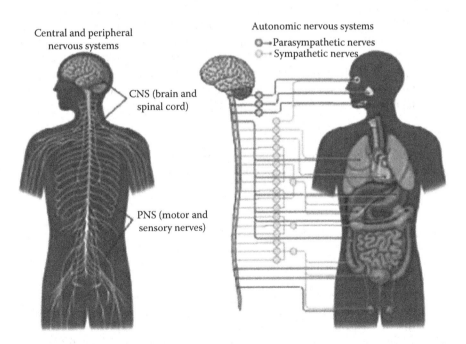

FIGURE 2.17 Peripheral nervous system. (From Gray, H., *Anatomy of the Human Body*, Lea & Febiger, Philadelphia, PA, 1918, on-line edition published May 2000 by Bartleby.com; © 2000 Copyright Bartleby.com, Inc.)

The nerves we use to consciously control parts of the body are components of the somatic or voluntary nervous system. Although these function are considered voluntary, there is some autonomy associated with their behavior as they function automatically even in the event of a coma.

The somatic nervous system provides the human with the ability to react consciously to environmental changes and includes 31 pairs of spinal nerves and 12 pairs of cranial nerves. This system also controls the movements of skeletal muscles.

Thirty-one pairs of spinal nerves extend from various segments of the spinal cord. Each spinal nerve has a dorsal root and a ventral root. The dorsal root contains afferent (sensory) fibers that transmit information to the spinal cord from the sensory receptors. The ventral root contains efferent (motor) fibers that carry messages from the spinal cord to the effectors.

From the sensory receptors, 12 pairs of cranial nerves transmit information relative to the human senses, including those of balance, smell, sight, taste, and hearing. This information is processed in the CNS and the resulting orders are transmitted back through the cranial nerves to the skeletal muscles that control movements.

The involuntary or autonomic nervous system maintains the body's equilibrium or homeostasis. As its name implies, this system works automatically and without voluntary input. It contains receptors within internal organs, the afferent nerves that relay the information to the CNS, and the efferent nerves that relay the action back to the effectors. The effectors in this system are smooth muscle, cardiac muscle and glands, and all structures that function without conscious control. An example of autonomic control is the movement of food through the digestive tract.

The efferent portion of the autonomic system is further divided into sympathetic and parasympathetic systems. The sympathetic nerves mobilize energy for the "fight-or-flight" reaction when the body experiences physical or psychological stress, causing increased blood pressure, breathing rate, and blood flow. Conversely, the parasympathetic nerves have a calming effect that slows the heartbeat and breathing rate to support other body functions such as digestion and elimination.

2.4.5.2.1   Reflexes

The relationship between sensory and motor neurons can be seen in a reflex (rapid motor response to a stimulus). Reflex arcs have five basic elements:

- receptor
- sensory neuron
- integration center (CNS)
- motor neuron
- effector

Reflexes are quick because they involve few neurons and are either somatic (resulting in contraction of skeletal muscle) or autonomic (activation of smooth and cardiac muscles).

Spinal reflexes are somatic reactions mediated by the spinal cord. In a spinal reflex, the messages are sent almost simultaneously to the spinal cord and brain. The reflex triggers the response without waiting for brain analysis or feedback. For example, if a finger touches something hot, the finger jerks away from the danger. The burning sensation becomes an impulse in the sensory neurons when the body perceives the heat and pain associated with the external stimuli. These sensory and motor neurons synapse (or connect) in the spinal cord and produce the fast response of pulling the finger away.

2.4.6 INTEGUMENTARY (SKIN) SYSTEM

The integumentary system is composed of various parts, including the skin, hair, nails, and sweat glands.

2.4.6.1 Skin

The skin is the largest organ of the body with a surface area of about $2\,m^2$ and in most places it is no more than $2\,mm$ thick; however, the mass of the skin exceeds that of all other organs. The integumentary system acts as a protective barrier and is the first line of defense for the body. The primary functions of the system are to maintain body temperature, protect internal organs, eliminate wastes, protect from diseases, and retain bodily fluids. The skin is also an organ of sensory perception experiencing sensations of pain, touch, pressure, and thermal conditions. Three layers of tissue make up the skin; from the surface downward they are defined as the epidermis, dermis, and subcutaneous layer. The skin varies in thickness from $0.5\,mm$ on the eyelid to 3–$4\,mm$ on the palms of the hands and soles of the feet. The skin is subject to multiple occupational risks from heat, cold, moisture, radiation, germs, and penetrating objects. Surveys indicate that dermatological disorders are second only to injuries as occupational incidents in the American workplace (Taylor, 2001) and thus occupational ergonomist and safety professionals should understand how to mitigate risk to this vitally important organ (Figure 2.18).

2.4.6.2 Hair and Nails

The hair and nails are modified forms of skin cells formed by a process that changes living epidermal cells into dead cells. Keratin is the major structural element in skin and nails. The hair and nails also provide protection (i.e., nose hair, toe nails) and the growth rate of both depends on many factors, including age, health, and nutrition.

2.4.6.3 Glands

Two primary types of glands are located in the dermis. The first is the sweat gland which produces perspiration and supports the activity of cooling the body when heated. The second is the sebaceous or oil gland, usually located in or near hair follicles. These glands are located in all parts of the body with the exception of the palms and soles. The dermis of the face and scalp has numerous sebaceous glands.

2.5 SUMMARY

The systems of the body work simultaneously to allow the body to function and perform automatic and voluntary activities. The sensory capabilities that provide the external stimuli to the systems of the body are covered in Chapter 3. The succeeding

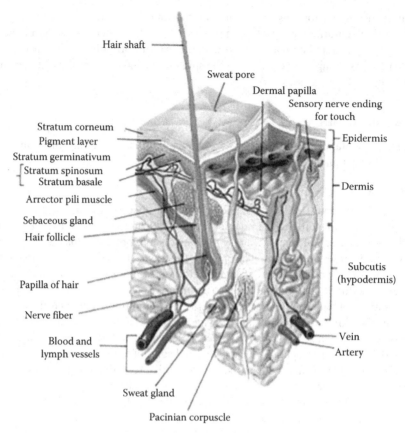

Hair shaft

Sweat pore
Dermal papilla
Sensory nerve ending
for touch

Stratum corneum
Pigment layer
Stratum germinativum
Stratum spinosum
Stratum basale
Arrector pili muscle
Sebaceous gland
Hair follicle

Epidermis

Dermis

Papilla of hair

Nerve fiber

Blood and
lymph vessels

Subcutis
(hypodermis)

Vein
Artery

Sweat gland
Pacinian corpuscle

FIGURE 2.18 Skin. (From Gray, H., *Anatomy of the Human Body*, Lea & Febiger, Philadelphia, PA, 1918, on-line edition published May 2000 by Bartleby.com; © 2000 Copyright Bartleby.com, Inc.)

chapters will discuss the impact of the occupational environment on these systems as well as risk mitigation approaches to protect the systems when necessary (Figure 2.19).

Case Study

Cell Phone Safety: Study to Probe Mobile Health Risk
By Matt McGrath
Science reporter, BBC World Service

The world's largest study on the safety of using mobile phones has been launched by researchers in London. The project will recruit 250,000 phone users across five different European countries including the UK. It will last between 20 and 30 years and aims to provide definitive answers on the health impacts of mobile phones. Research to date has shown no ill effect, but scientists say those studies may be too short to detect longer term cancers and other

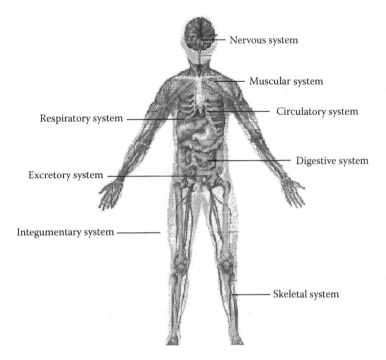

FIGURE 2.19 Select systems of the human body.

diseases. The study is known as Cosmos—the cohort study on mobile communications. It is being funded in the UK by the Mobile Telecommunications and Health Research Programme, an independent body, for an initial five year period.

New uses

A member of that group, Professor Lawrie Challis, said the study was crucial. "We still cannot rule out the possibility that mobile phone use causes cancer. The balance of evidence suggests that it does not, but we need to be sure."

The co-principal investigator of the study, Dr Mireille Toledano from Imperial College London, added that there are still "gaps in our knowledge, there are uncertainties."

She said: "The best thing we can do as a society is to start now to monitor the health of a large number of users over a long period of time—that way we can build up a valuable picture as to whether or not there are any links in the longer term." She stressed the study was not just about brain cancer. People were now using mobiles in many different ways including surfing the web, which means the phones are not always held against the head. She added: "We will be looking at a range of different health outcomes, including other forms of cancers such as skin cancers and other brain disease such as neurodegenerative diseases. We will also be monitoring things like if there's a change in the frequency of symptoms such as headaches, tinnitus, depression or sleep disorders. These are things that people commonly report in association with their mobiles and these are things we are going to be following up on over time as well."

Bias

One of the greatest concerns about research to date is that it has usually depended on participants recalling how much they have used their phones. Scientists say this can affect the outcome. The Cosmos project will be prospective—meaning that it will record actual phone use into the future. Around 100,000 mobile phone users in the UK across different networks will be invited to take part. Mobile phone users will also be recruited in Finland, Denmark, Sweden and the Netherlands. Dr Toledano says the scientists will monitor mobile usage but not the numbers people call. And once participants fill in a questionnaire and give permission to access their records the project will operate very much in the background. "It's really not intrusive," she said. "Most of it is very passively done, once they've given us their permission to sign up, we've really made it very easy and that's why we'd encourage people to take part." The researchers will report their initial findings in 5 years. They will monitor WIFI, cordless phones and the use of baby monitors by participants as well as mobiles, to obtain a complete picture of exposure to all types of electromagnetic radiation. More than 70 million phones are in use in the UK at present, out of a global total of 5 billion.

Source: Total contents of Case Study reprinted with permission from Ergoweb.com

EXERCISES

2.1 Explain each of the systems of the body.

2.2 Describe the structure of the spinal column.

2.3 What are the primary components of an intervertebral disc?

2.4 What is a motor unit?

2.5 What is a synovial joint?

2.6 What aspect of the muscular system is impacted through repetitive heavy lifting?

2.7 For each of the systems discussed in this chapter, identify an occupational task that will impact that system. Explain how.

2.8 Define a task (activities, equipment, and duration) that will impact the circulatory system.

2.9 What is the largest system of the body? What are the primary occupational risk factors associated with this system?

2.10 Define a task activity and explain how at least three systems of the body are simultaneously impacted by task performance.

REFERENCES

Gascoigne, B. (2001). History of anatomy, HistoryWorld (From 2001, ongoing). http://www.historyworld.net/wrldhis/PlainTextHistories.asp?groupid=44&HistoryID=aa05

Gray, H. (1918). *Anatomy of the Human Body*, Lea & Febiger: Philadelphia, PA. On-line edition published May 2000 by Bartleby.com; © 2000 Copyright Bartleby.com, Inc.

Huffman, C. (2008). Alcmaeon, *The Stanford Encyclopedia of Philosophy* (Ed. E.N. Zalta), Winter 2008 edn., http://plato.stanford.edu/archives/win2008/entries/alcmaeon/

The Visual Dictionary, Joints of the Human Body, http://www.infovisual.info/03/026_en.html (accessed 2011).

3 Senses of the Human Body and Measurement of Environmental Factors

3.1 LEARNING GOALS

This chapter will provide the student with knowledge of the human senses and the sensory process for each system. A discussion of tools and techniques to assess the sensory impact is offered to teach the process to evaluate the environment as it relates to the senses. Measurement and management of the occupational conditions on the human senses will also be addressed.

3.2 KEY TOPICS

An introduction to human sensory functions and measurement of the level of sensory capability is the focus of this chapter. Guidelines used to promote ergonomic design that is compatible with the human senses are covered, as well as a discussion of know occupational risk factors.

3.3 INTRODUCTION AND BACKGROUND

Much of what we know about how smells, sights, tastes, textures, and sounds are processed in the brain is due to the work of Wilder Penfield and colleagues during the mid-twentieth century. By directly stimulating different areas of exposed cortical tissue during neurosurgery, these doctors were able to map the specific areas of the brain that are activated as a result of different types of sense stimulation. While the five senses are critical to the brain's ability to process information about the natural world, recent knowledge affirms that they are vital to mental health, and the loss of one or more of these sensations can lead to physical and emotional distress (http://www.sciencedirect.com.proxy.library.emory.edu/science/journal/00016918, *Acta Psychologica*, April 2001).

According to *Webster's Online Dictionary*, "sensory deprivation" is defined as "a form of psychological torture inflicted by depriving the victim of all sensory input." In fact, depriving human beings of their natural human senses has been officially considered torture since the European Court of Human Rights ruled in 1978 (*Ireland v. United Kingdom*) that the use of deprivation techniques such as wall-standing, hooding, subjection to noise, sleep deprivation, and deprivation of food and drink were inhumane. The media took a great interest in the subject surrounding this landmark case as well. For instance, the prominent 1960s' television show *The Twilight Zone* featured an episode where an astronaut isolates himself in a sensory deprivation room in order to mimic the conditions of being on the moon. "Where is Everybody?", as the first episode of the twilight series was titled, results in the astronaut's mental deterioration as a result of his isolation (http://www.websters-online-dictionary.org).

This episode is an example of a growing concern about the effect of environmental factors, like isolation from others and deprivation of sensory stimulation, on the mind, and more specifically, human task performance. Far from the Industrial Revolution era fixation on working employees for the longest hours at the lowest wages possible, today's businesses and manufacturers are looking for ways to improve the quality of work rather than simply increasing the quantity of mechanical output. For instance, over the last 30 years, millions of dollars have been invested in sleep research, to investigate the amount of sleep necessary for adequate performance, as lack of sleep diminishes sensory alertness and brain awareness. In one such study, three groups of healthy subjects were randomly assigned various lengths of post-lunch naps, and tested in areas of alertness, task performance, and autonomic balance. The results of this study, and others like it, demonstrated the benefits of a short post-lunch nap on overall performance quality, and as a result many businesses are trying to incorporate napping facilities for the use of their employees (*European Journal of Applied Physiology and Occupational Physiology*).

3.4 SENSORY FUNCTIONS

The various systems of the body interact to create sensory functions. The human sensory functions are the means by which information is perceived from the external environment by humans. In the absence of sensory capabilities, the body would be cut off from the world as no knowledge of the external environment could be received. The primary senses in ergonomic study include

- Vision
- Audition
- Cutaneous
- Olfactory
- Gustation

The senses are essentially receptors, constantly receiving input from the environment. Sensory functions allow the body to respond to stimuli. Nerves and sensory receptors respond to both internal and external stimuli. These stimuli are received by nerves and receptor cells and are converted into impulses that travel to the brain, where the signals are processed. In addition to the traditionally considered "five senses"—sight, hearing, smell, taste, and touch—humans have senses of motion (kinesthetic sense), heat, cold, pressure, pain, and balance (New World Encyclopedia, 2008).

3.4.1 VISION

Vision is a complicated process that requires numerous components of the human eye and brain to work together. The eye is the primary organ of the visual sense. The

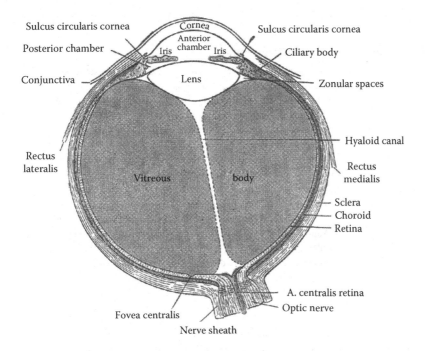

FIGURE 3.1   Eye. (From Gray, H., *Anatomy of the Human Body*, 20th edn., Lea & Febiger, Philadelphia, PA, 1918, retrieved September 9, 2009, from http://www.bartleby.com/107/)

eye is connected to the brain through the optic nerve and is a complex structure that contains a transparent lens. This lens focuses images and light that enter through the pupil onto the retina. The retina is covered with two types of light-sensitive cells known as rods and cones. The cone cells are sensitive to color and are located in the part of the retina called the fovea, where the light is focused by the lens. The rod cells are not sensitive to color, but have greater sensitivity to light than the cone cells. These cells are located around the fovea and are responsible for peripheral vision and night vision. After the photoreceptors collect the light and send the signals to a network of neurons, electrical impulses are generated and subsequently transmitted to the brain through the optic nerve (Figure 3.1; Table 3.1).

3.4.1.1   Visual Fatigue

The reliance on the vision as the primary sensory mode can lead to overuse in the occupational environment, especially in tasks that utilize any type of video display terminals. "Visual fatigue" is the term used to describe conditions experienced by individuals when the visual sense is stressed due to incompatibility of the occupational environment with the visual system. The effects of visual fatigue may include loss of productivity, decline in quality, increased human error, increased accident rate, visual complaints, and long-term discomfort (Grandjean, 1997). To minimize visual fatigue, the occupational environment and task requirements should be designed with consideration of visual capabilities, limitations as well as lighting

TABLE 3.1
Visual Sense: Key Terms in Ergonomics

Visual field	The area, measured in degrees, which can be seen by both fixated eyes
Accommodation	The action of focusing on targets at various distances
	• Near-point is the closest distance that an object can be focused on without conscious accommodation.
	• Far-point is the farthest point that can be focused without conscious accommodation.
	• Amplitude of accommodation: the difference between the near-point and far-point.
Convergence	• The ability of the eyes to compensate for the difference the angle formed by the difference in the lines of sight from each when focusing on a target.
	• If targets are too close to the eye then convergence can become difficult.
Adaptation	• Adjustments that the pupils of the eyes make to changes in illumination are called adaptation.
	• Adaptation from light to dark takes about 30 min. Adaptation from dark to light happens much more quickly, normally within a minute.

guidelines. Understanding and controlling glare is also critical in minimily risk of visual fatigue. Methods to control direct and indirect glare are listed in Table 3.2 (Kaufman and Christensen, 1972; Morgan et al., 1963).

These methods also include considering such factors as the size of visual targets, distance and angle of the task visual demands, lighting, glare, contrast, time allowed, biomechanics of the task, and environmental conditions.

TABLE 3.2
Techniques for Controlling Glare

To Control Direct Glare	To Control Indirect Glare (Veiling Reflections and Reflected Glare)
Position luminaires, the lighting units, as far from the operator's line of sight as is practical	Avoid placing luminaires in the indirect glare offending zone
Use several low-intensity luminaires instead of one bright one	Use luminaires with diffusing or polarizing lenses
Use luminaires that produce a batwing light distribution, and position workers so that the highest light level comes from the sides, not front and back	Use surfaces that diffuse light, such as flat paint, non-gloss paper, and textured finishes
Use luminaires with louvers or prismatic lenses	Change the orientation of a workplace, task, viewing angle, or viewing direction until maximum visibility is achieved
Use indirect lighting	
Use light shields, hoods, and visors at the workplace if other methods are impractical	

Source: Adapted from Kaufman and Christensen, 1972; Morgan et al., 1963; Eastman Kodak, 1986.

3.4.2 AUDITION

Hearing occurs when sound waves emitted from a source are collected by the outer ear and channeled along the auditory canal to the eardrum, which vibrates according to the frequency and intensity of the arriving sound wave. The science of sound is known as acoustics, and includes sound production, transmission, and processing. The ear consists of three components where each contain numerous small parts (Box 3.1).

- The external or outer ear consists of
  - Pinna or auricle—the outside part of the ear.
  - External auditory canal or tube—the tube that connects the outer ear to the inside or middle ear.
  - Tympanic membrane—also called the eardrum. The tympanic membrane divides the external ear from the middle ear.
- The middle ear (tympanic cavity) consists of
  - Ossicles—three small bones that are connected and transmit the sound waves to the inner ear. The bones are called malleus, incus, and stapes.
  - Eustachian tube—canal that links the middle ear with the throat area.
- The inner ear consists of
  - Cochlea—contains the nerves for hearing.
  - Vestibule—contains receptors for balance.
  - Semicircular canals—contain receptors for balance.

3.4.2.1 How Do We Hear?

Sound is any vibration that stimulates an auditory sensation. When a sound is made outside the outer ear, the sound waves, or vibrations, travel down the external auditory canal and strike the eardrum (tympanic membrane). The eardrum vibrates. The vibrations are then passed to three tiny bones in the middle ear called the ossicles. The ossicles amplify the sound and send the sound waves to the inner ear and into the fluid-filled hearing organ (cochlea). Once the sound waves reach the inner ear, they are converted into electrical impulses by small, sensitive hair cells, which the auditory nerve sends to the brain. The brain then translates these electrical impulses as sound.

**BOX 3.1 HEARING IS BASED ALMOST
ENTIRELY UPON PHYSICAL MOVEMENT**

Hearing is based entirely on movement, unlike the many other chemical-based senses of the body.

The purpose of the ear's structure is to provide a path for information, which culminates at the ear drum.

TABLE 3.3

Speech Intensity Table

Description	Typical Intensity of Source, dB(A?)	Distance	Message Characteristic
Soft whisper		3–6 in.	Secret communication
Audible whisper	44–69	8–20 in.	Confidential or personal
Normal voice	50–75	20–60 in.	—
Loud voice	56–81	5–8 ft	Nonpersonal, intermediate group
Very loud voice	62–87	To 20 ft	Group address
Shouting	68–93	Upper limit	Hailing or emergency communication

Source: Woodson, W.E., Tillman, B., and Tillman, P. *Human Factors Design Handbook*, 2nd edn., McGraw-Hill, Inc., New York.

In an occupational environment many factors impact the quality and consistency of hearing, including the individual's hearing ability, background noise, competing auditory signals, and environmental factors. The level (or intensity) of the auditory message, as well as the factors listed earlier, should be taken into consideration when designing the distance from which the message should be transmitted. Table 3.3 shows typical speech intensity levels for common communication (Woodson, 1992).

The application of sound auditory guidelines and principles can be effective in reducing negative effects of excessive noise. Although individual perceptions of noise may vary, excessive noise levels that exceed the Occupational Safety and Health Administration (OSHA) guidelines can significantly impact task performance. Table 3.4 provides a summary of some conditions that can result from exposure to excessive noise levels (Woodson, 1992).

3.4.2.2 Noise-Induced Hearing Loss

Noise is any acoustic phenomenon that annoys the listener, making noise a psychological and subjective issue. Noise at safe levels can be annoying and distracting but will not physically affect hearing. However, when an individual is exposed to harmful noise—sounds that are too loud or loud sounds that last a long time—the sensitive structures in the inner ear can be temporarily or permanently damaged, resulting in noise-induced hearing loss (NIHL). Additionally high noise levels can adversely impact task performance. Table 3.3 provides a summary of the impact various sound levels on performance (Woodson, 1992). The small, sensory hair cells convert sound energy into electrical signals that travel to the brain. Once damaged, the sensory hair cells cannot grow back.

Risk factors for NIHL are found in both the occupational and recreational environments. NIHL is completely preventable given good care of the hearing sense. All individuals should understand the hazards of noise and follow safe hearing practices in everyday life. To protect hearing in the occupational environment consider the guidelines provided in Table 3.5.

TABLE 3.4
Noise Effects on Performance

Noise Level, dB	Effects
150	Reduced visual acuity, chest-wall vibration, changes in respiratory rhythm, and a "gagging" sensation.
120	Loss of equilibrium.
110	Chronic fatigue and digestive disorders.
105	Reduced visual acuity, stereoscopic acuity, and near-point accommodation and permanent hearing loss when exposure continues over a long period (months).
100	Serious reduction in alertness. Attention lapses occur, although attention duration is usually not affected. Temporary hearing loss occurs if no protection is provided in the region 600–1200 Hz. Most people will consider this level unacceptable, and 8 h is the maximum duration they will accept.
95	Considered to be the upper acceptance level for occupied areas where people expect the environment to be noisy. Temporary hearing loss often occurs in the range of 300–1200 Hz. Speech will be extremely difficult, and people will be required to shout, even though they may be talking directly into a listener's ear.
90	At least half of the people in any given group will judge the environment as being too noisy, even though they expected a noisy environment. Some temporary hearing loss in the range of 300–1200 Hz occurs. Skill errors and mental decrements will be frequent. The annoyance factor is high, and certain physiological changes often occur (e.g., the pupils dilate, the blood pressure increases, and the stroke volume of the heart may decrease). Listening to a radio is impossible without good earphones. The maximum duration that most people will accept is 8 h.
85	The upper acceptance level (noise expected) in the range of 150–1200 Hz. This is considered the upper comfort level, although some cognitive performance decrement can be expected, especially where decision making is necessary.
80	Conversation is difficult (i.e., people have to converse in a loud voice less than 1 ft apart). It is difficult to think clearly after about 1 h. There may be some stomach contraction and an increase in metabolic rage. Strong complaints can be expected from those exposed to this level in confined spaces, and 8 h is the maximum.
75	Too noisy for adequate telephone conversation. A raised voice is required for conversants 2 ft apart. Most people will still judge the environment as being too noisy.
70	The upper level for normal conversation, even when conversants are close together (at a distance of 6 ft people will have to shout). Although persons such as industrial workers and shipboard personnel who are used to working in a noisy environment will accept this noise level, unprotected telephone conversation will be difficult (upper phone level is 68 dB).
65	The acceptance level when people expect a generally noisy environment. Intermittent personal conversation is acceptable. About half of the people in a given population will experience difficulty sleeping.
60	The upper limit for spaces used for dining, social conversation, and sedentary recreational activities. Most people will rate the environment as "good" for general daytime living conditions.

(continued)

TABLE 3.4 (continued)
Noise Effects on Performance

Noise Level, dB	Effects
55	The upper acceptance level for spaces where quiet is expected (150–2400 Hz). People will have to raise their voices slightly to converse over distances greater than 8 ft. This level of noise will awaken about half of a given population about half of the time. It is still annoying to people who are especially sensitive to noise.
50	Acceptable to most people where quiet is expected. About 25% will be awakened or delayed in falling asleep. Normal conversation is possible at distances up to 8 ft.
40	Very acceptable to all. The recommended upper level for quiet living spaces, although a few people may still have sleep problems.
30	Necessary for specialized listening tasks (e.g., threshold signal detection).
<30	Introduces additional problems; that is, low-level intermittent sounds become disturbing. Some people have difficulty getting used to the extreme quiet, and a few may become psychologically disturbed.

Source: Woodson, W.E., Tillman, B., and Tillman, P. *Human Factors Design Handbook*, 2nd edn., McGraw-Hill, Inc., New York.

NIHL is a commonly occurring condition in many occupations. Strategies to reduce the risk of occupational NIHL include the following:

- Reduce or avoid the generation of the sound by designing machine parts to minimize noise production.
- Impede the transmission of noise by using barriers, encapsulating noise source, and using sound absorbing surfaces.

The nature of the noise source may require adapting standard procedures or creating unique approaches for reducing the impact of noise. A combination of engineering controls, administrative controls, and personal protective equipment should be used to address hazardous noise and reduce the likelihood of NIHL (Box 3.2).

3.4.2.3 Current Technology: iPods and MP3(4) Impact on Hearing

The rapid development of digital technology has resulted in new kinds of portable music players, including MP3 players such as iPods, in which sound quality is no longer distorted at higher volumes. Because they are equipped with improved headphones, sound-leakage is minimal, allowing these devices to be used at high volume levels in most environments without disturbing others. As a result, the conditions for higher sound levels and longer exposure times are created, both of which are known to increase the risk of hearing damage. MP3 players, thus, may be the most important risk factor for music-induced hearing loss in young people (Vogle, 2008).

TABLE 3.5
Upper Noise Level Limits

Sound Level, dBA	Type of Activity	Communication Equipment	Office Application
108	Maximum design limit for AMC equipment (hearing protection required)	No direct communication	Not recommended
100	Armored vehicles (hearing protection required)	Electrically aided communication satisfactory with attenuating helmet or headset; limited shouted communication possible with difficulty	Not recommended
90	Material that is beyond the state of the art of meeting 85 dBA (hearing protection required)	Shouted communication possible at short distances (1–2 ft)	Not recommended
85	Acceptable level for unprotected hearing for 8 hr exposures	Shouted communication possible at several feet (3–4 ft); telephone use difficult	Not recommended
75	Maintenance shops, garages, and keypunch areas	Occasional telephone use and occasional direct communication at up to 5 ft is acceptable	Not recommended
65	Operation centers, mobile command and communication centers, computer rooms, word processing centers, kitchens, and laundries	Frequent telephone use and frequent direct communication at up to 5 ft is acceptable	Business machine offices
55	Drafting rooms, laboratories, and conferences with two or three people	No difficulty with telephone use and occasional direct communication at up to 15 ft	Shop offices and general secretarial areas
45	Libraries, conference rooms, command and control centers, theatres, and sleeping areas	No difficulty with direct communication	General offices
35	Recording studios and large conference rooms	Areas requiring unusually extreme quiet	Executive offices

Source: Woodson, W.E., Tillman, B., and Tillman, P. *Human Factors Design Handbook*, 2nd edn., McGraw-Hill, Inc., New York.

Given the international popularity of portable music devices, exposure to unsafe sound levels has increased dramatically, and millions of adolescents and young adults are experiencing short-term hearing loss and are potentially at risk of permanent hearing loss through listening to their favorite music. Music-induced hearing loss may be evolving into a significant social and public health problem. Increasing

BOX 3.2   OLFACTORY JOBS

- Culinarian (cook, chef)
- Enologist (wine maker/taster)
- Perfumer
- Quality assurance (food and beverage companies)
- Safety inspector (chemical/hazard, smoke detection)
- Law enforcement (drug and alcohol detection)
- Medical workers
- Firefighters (gas/chemical leaks, smoke detection)

numbers of adolescents and young adults already show related symptoms, such as distortion, tinnitus, hyperacusis, and threshold shifts.

3.4.3 OLFACTION

Olfaction involves the sense of smell, and the major organ associated with this sense is the nose. In the upper part of the human nostril, several million smell-reacting sensors comprise the olfactory epithelium, which is responsible for the detection of smell (Figure 3.2). These smell receptors are connected to the olfactory nerve and transmit the sensations to the brain (Myers, 2004). These receptors allow the human to detect as many as 10,000 different smells. However, this is small compared to olfactory capabilities of dogs. Dogs have a sense of smell that is thousands of times more sensitive than humans, as they have 25 times more olfactory receptors than humans (Correa, 2005).

The sense of smell has physiological impacts that stimulate the nervous system, and psychological and psychogenic impacts that effect one's attitude, mood, and ability to perform. The sale of candles, air fresheners, and other fragrances has grown substantially in recent years. The multibillion dollar a year aromatherapy industry is built on the olfactory sense (Figure 3.3) (Demand Studios, 2009).

Several techniques are used to describe and distinguish odor qualities including those developed by Fanger (1988). These two measures are the olf (from the Latin word *olfactus* = olfactory sense) and the decipol (from the Latin "pollution" = pollution) (Konz, 1997). One olf is the emission rate of air pollutants from a standard person in an office or nonindustrial environment. Likewise, one decipol is the pollution caused by 1 olf ventilated by 10 L/s of unpolluted air (Konz, 1997). Table 3.6 shows typical odor perception for sample olf values.

Seven classes of smell are widely recognized: ethereal, camphoraceous, musky, floral, minty, pungent, and putrid. The Henning's Smell Prism (Iowa State University, 2004) presents six odors at the corners of a prism allowing better visualization of combinations of scents that produce complex odors. The Henning's Smell Prism classifies odors as follows (Figure 3.4):

- Fragrant
- Putrid

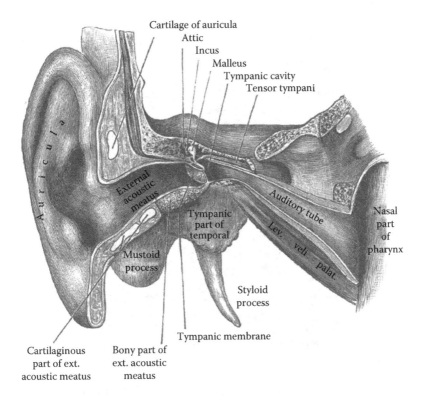

Cartilage of auricula
Attic
Incus
Malleus
Tympanic cavity
Tensor tympani

Auricula

External acoustic meatus

Tympanic part of temporal

Auditory tube

Lev. veli palat.

Nasal part of pharynx

Mastoid process

Styloid process

Cartilaginous part of ext. acoustic meatus

Bony part of ext. acoustic meatus

Tympanic membrane

FIGURE 3.2 Ear. (From Gray, H., *Anatomy of the Human Body*, 20th edn., Lea & Febiger, Philadelphia, PA, 1918, retrieved September 9, 2009, from http://www.bartleby.com/107/)

- Ethereal
- Spicy
- Resinous
- Burned

While the classes of odors previously described are recognized as the seven primary classes of stimulants, most odors are considered complex odors encompassing two or more olfactory receptors. The intensity of the odor is another factor that can pose challenges to categorizing odors. Psychological studies have indicated the olfactory sense is the sensory function most strongly associated with memory recall. Dr. Jay Gofried at University College London states "Our study suggests that, rather than clumping together the sights, sounds, and smells of a memory into one bit of the brain, the memory is distributed across different areas and can be re-awakened through just one of our sensory channels" (Gottfried et al., 2004).

Different scents have also been associated with performance in a number of areas. A 2007 German study showed that participants who smelled a rose scent while learning and then again while experiencing Rapid Eye Movement (REM) sleep performed better on a quiz related to the material they had learned the previous day (Hitti, 2007).

FIGURE 3.3 Aromatherapy industry products. (From Demand Studios, 2009, retrieved September 9, 2009, from http://cdn-write.demandstudios.com/upload/3000/200/00/0/33200.jpg)

TABLE 3.6
Odor Perception

Olf Value	Source
0–5	Per square meter of materials in the office
1	Sedentary person, 1 met
5	Active person, 4 met
6	Smoker, average
11	Active person, 6 met
25	Smoker, when smoking

The inhaling of peppermint vapor has been proven to improve an athlete's speed, strength, and endurance and substantially decrease fatigue (Raudenbush, 2001). The emerging field of aromachology, or the study of aromas and how they affect human behavior and performance, is providing promising possibilities in areas as diverse as athletic performance and improved memory.

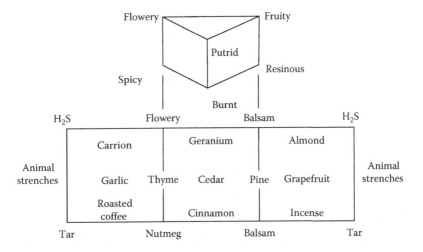

FIGURE 3.4 Henning's smell prism. (From Iowa State University, The science of smell part 2: Odor chemistry, 2004, retrieved January 23, 2011, from http://www.extension.iastate.edu/ Publications/PM1963B.pdf)

3.4.4 GUSTATION

Gustation is the sense of taste. This sense overlaps and depends largely on the senses of smell and tactation. In other words, the smell and texture of objects that the tongue engages impact the perception and degree of favorability of the item. When considering the sense of taste alone, the receptors for taste or taste buds are used to assess the type of taste within substances. Nearly 10,000 taste buds are located primarily on the human tongue but also on the roof of the mouth and near the pharynx. They are able to detect four basic tastes: salty, sweet, bitter, and sour. The taste buds are continuously replaced and are renewed approximately every 2 weeks. Different parts of the tongue are more prone to detect certain tastes, such as the tip of the tongue, which is most sensitive to sweets (Smith and Margolskee, 2001).

Given the strong dependence on the olfactory sense, approximately 75% of what is perceived as taste actually comes from the sense of smell. Taste buds allow us to perceive only bitter, salty, sweet, and sour flavors, but it is the odor molecules from food that provide most of the taste sensation. When food is placed in the mouth, odor molecules travel through the passage between the nose and mouth to olfactory receptors cells at the top of the nasal cavity located just beneath the brain and behind the bridge of the nose. This information is interpreted as a contributing aspect of the perceived taste of the food.

Gustation is not used extensively in the occupational environment, but there are certain occupations that benefit from one's sense of taste. Occupations that rely on the olfactory sense include wine tasters, food critics, and food samplers (Figure 3.5).

FIGURE 3.5   Primary taste sensitivity on human tongue.

3.4.5 TACTATION

Tactation, or the sense of touch is distributed throughout the body. Nerve endings in the skin and other parts of the body transmit sensations to the brain. Some parts of the body have a larger number of nerve endings and, therefore, are more sensitive. Four kinds of touch sensations can be identified: thermal (cold or heat and moisture), pressure, electronic stimulation, and pain.

Sensory capabilities located in the skin are referred to as cutaneous and comprise the tactation sense. The four categories of cutaneous senses include

- Mechanoreceptors that sense tactation, that is, contact touch and pressure
- Thermoreceptors that sense warmth or cold relative to each other and the body's neutral temperature
- Electroreceptors that respond to electrical stimulation of the skin (theoretical, but not a proven factor)
- Nocireceptors that sense pain

These receptors work individually and together to sense the level of tactation to the body.

3.5   ENVIRONMENTAL FACTORS IN ERGONOMICS

The environmental factors that influence the task performance and comfort of workers in the occupational environment can be classified in one or more of the following broad categories (Box 3.3):

BOX 3.3   KEY TERMS

- *Photometric energy*—radiant energy modified by the luminous efficiency of the standard observer.
- *Illuminance*—the amount of light falling on a surface (lx)
- *Luminance*—the amount of light energy reflected (or emitted from a surface)

- Visual factors: lighting
- Auditory factors: sound levels
- Thermal factors: temperature and humidity

These are further discussed in the following sections.

3.5.1 VISUAL FACTORS LIGHT LEVELS

The ability to see, particularly in an occupational setting, depends upon several environmental and workplace factors including: the level of illumination, the contrast, reflection of the work surface, and glare. Each of these factors must be considered and sound ergonomic principles must be utilized in workplace design for proper illumination. Key terms used to describe the resulting light levels include photometric energy, illumina and luminance. (Box 3.3). Discounting the importance of providing an appropriately illuminated work environment can result in OSHA fines, poor quality, reduced output, worker fatigue, job dissatisfaction, and increased risk of injury and subsequent liability. Discussions in the following sections rely on OSHA, the American Illuminating Engineering Society, and Woodson (1992).

3.5.1.1 Illumination

Light is defined as any radiation capable of causing a visual sensation. Objects are detected by the human eye in one of two methods, either by emitting a source of light, known as radiance, or by reflecting light, known as irradiance. Both radiance and irradiance are measured in watts per square meter (W/m^2). Illumination is the measure of the stream of light falling on a surface. This light may come from the sun, lamps, or any other bright source. Illumination is measured in lux or foot candle (fc) and is the light density per square foot.

$$1 \text{ lx} = 1 \text{ lumen per square meter} = 0.1 \text{ fc}$$

3.5.1.2 Engineering Procedures in the Design of the Environment for Proper Vision

Proper vision requires sufficient quantity and quality of illumination. Special requirements on visibility (especially with the diminished visual acuity of the elderly) require care in the arrangement and delivery of illumination. Color vision especially requires sufficient light. If an object or environment is well illuminated (above 0.1 lx), the human eye uses photopic vision for sensing the image. Proper color perception occurs under photopic conditions. Vision at illumination levels between 0.1 and 0.01 lx is known as mesopic. In mesopic, brighter areas appear in color, but darker areas are seen in shades of gray. If illuminance falls between 0.1 and 0.01 lx, both cones and rods respond creating a mesopic condition. In light conditions below 0.01 lx, vision is limited to scotopic vision, in which only shades of gray, black, and white can be detected. Table 3.7 provides functional ranges of the visual system and associated capabilities (Gavriel, 1997).

TABLE 3.7

The Functional Ranges of Visual System Capabilities

Name	Dominance Range (cd/m²)	Photoreceptor Active	Wavelength Range (nm)	Capabilities
Photopic	>3	Cones	380–760	Color vision Good detail discrimination
Scotopic	<0.001	Rods	380–760	No color vision Poor detail discrimination
Mesopic	>0.001 and <3	Cones and rods	380–760	Diminished color vision, reduced detail discrimination and a shift in spectral sensitivity as adaptation luminance moves from photopic to scotopic

Source: Gavriel, S. (Ed.): *Handbook of Human Factors and Ergonomics.* 1997. Copyright Wiley-VCH Verlag GmbH & Co. KGaA.

Luminance is the measure of light coming from a surface (a function of the light that is emitted or reflected from the surface of walls, furniture, and other surfaces) and is measured in units of candela per square meter (cd/m²). The most significant factor in detection of light is the luminance of an object, or the energy reflected or emitted from it. Luminance of an object is determined as a function of incident illuminance and reflectance. Reflectance is an inherent property of an object and is defined as the ratio of the amount of light reflected compared to the amount of light received by an object.

$$\text{Luminance} = \text{Illuminance} \times \text{reflectance} \times \pi^{-1}$$

For example, if a bright table surface has a reflectance of 70% and the incident light has an illumination figure of 400 lx, the luminance of the table will be 70% of $400/\pi = 89\,\text{cd/m}^2$ (Figure 3.6).

3.5.1.3   General Rules of Contrast

Vision is affected by contrast as well. Contrast is the difference between the luminance of an object and the luminance of surrounding surfaces, including its shadow. Contrast is calculated as follows:

$$\text{Contrast} = \left[\frac{(L_{max} - L_{min})}{L_{max}} \right] \times 100\%$$

where

L_{max} is the maximum luminance level of surrounding surfaces
L_{min} is the minimum luminance level of surrounding surfaces

FIGURE 3.6 Odor perception. (From Teamwork Photo & Digital, Sekonic, 2009, retrieved September 9, 2009, from http://www.teamworkphoto.com/images/sekonic/758d.jpg)

There are general rules that can be used when designing for appropriate levels of contrast. These are as follows:

- All of the objects and major surfaces in the visual field should be lit.
- Surfaces in the middle of the visual field should not have a contrast of more than 3:1.
- Contrast between the middle field and the edge of the visual field should not exceed 10:1.
- The working field should be brightest in the middle and darker toward the edges.
- Avoid excessive contrast to the sides and below the visual field.
- Light sources should not contrast with their background by more than 20:1.
- The maximum permissible range of contrast between items within a room is 40:1.

Effective use of contrast can be a useful means to communicate information and differentiate between task elements. Utilizing these guidelines will result in an appropriately contrasted environment.

3.5.1.4   Glare

To avoid unwanted or excessive glare, appropriate placement and indirect light sources can be used. Direct glare occurs when light meets the eye directly from a light source such as the headlights of an oncoming car. Direct glare can be reduced by placing high-intensity light sources outside the cone of 60° around the line of sight or by using several low-intensity light sources instead of one intense source.

Indirect glare is reflected from a surface into the eyes such as headlights of a car in your rearview mirror. Altering the angle of the indirect light source can be useful in eliminating indirect glare. Additionally, shields, hoods, and visors can be strategically placed around reflecting surfaces to keep reflected light out of the eye.

3.5.1.5 Sources and Measurement

As previously stated, Illumination, the measure of the stream of light falling on a surface, may come from the sun, lamps, or other bright sources. The level of illumination is measured in lux or foot candle (fc):

$$1 \text{ lx} = 1 \text{ lumen per square meter} = 0.1 \text{ fc}$$

Luminance is the measure of the brightness of a surface, where the perception of brightness of a surface is proportional to its luminance. Luminance is the measure of light coming from a surface (a function of the light that is emitted or reflected from the surface of walls, furniture, and other surfaces) and is measured in units of candela per square meter (cd/m^2).

A comparison of the luminance of various surfaces can be expressed as reflectance, which is the ratio between the total amount of light reflected by a surface compared to the light incident to the surface.

3.5.1.5.1 Light Sources

The type of lighting source used in an environment is dependent upon the lighting requirements, environmental factors, and preferences. The objective in lighting a room or workplace should be to accomplish the following:

- Suitable level of illumination—see Illuminating Engineering Society of North America (IESNA) guidelines for specific lighting design guidelines
- Spatial balance of surface's luminance—avoid sharp changes in luminance levels
- Temporal uniformity of lighting—focusing on bright and dark objects quickly
- Avoidance of glare
- Accommodation to the characteristics of the employee population
- Energy efficiency

The IES guidelines are widely accepted as a resource for designing occupational illumination levels to be compatible with task requirements. A variety of guidelines support task designers in understanding the impact of different levels of illumination on task performance. Table 3.8 provides a summary of these guidelines given the desired "alertness-level" requirements for a task.

TABLE 3.8

American Illuminating Engineering Society Guidelines

Alertness-Level Requirement	Average Light Level, fL
Maximum mental alertness required for highly complex mental task performance	50 and above
Medium mental alertness required for routine manual tasks, leisure reading, and/or stimulating social activity	40
Minimum mental alertness required for nondemanding social intercourse and/or perceptual-motor performance (dining, dressing, personal hygiene, etc.)	30
Rest, mental alertness for minimum interaction with other people (e.g., as in a bar or private dining environment)	15
Sleep	<3

Source: Woodson, W.E., Tillman, B., and Tillman, P. *Human Factors Design Handbook*, 2nd edn., McGraw-Hill, Inc., New York.

Some examples of lighting appropriate for use in occupational environments include filament lamps and fluorescent lights. Filament lamps are relatively rich in red and yellow rays. When used above a workplace they emit heat, and even with a lamp shade, a filament lamp can exceed temperatures of 60°C and can lead to discomfort and headaches. Fluorescent tubes are much more efficient than heated filament and do not produce the same levels of heat. The inside of the tube is covered with a fluorescent substance that converts the ultraviolet rays of the discharge into visible light, the color of which can be controlled by the chemical composition of the fluorescent material.

3.5.1.5.2   Measurement of Illumination

The IESNA has guidelines for lighting requirements and generally the more intense and delicate the work, the more lighting needed. There are several ways to measure light or illumination, and these methods have led to much confusion. The confusion stems from a combination of several sources creating multiple guidelines and the differences between physically described and humanly perceived units. Measurement of the quantity of radiant energy can be made in several ways, but the fundamental types of radiometry energy measurement include the following:

- Total radiant energy emitted from a source, per unit of time (radiant flux, expressed in watts)
- The energy emitted from a point in a given direction (radiant intensity, expressed in watts/steradian)
- The energy arriving at a surface at some distance from a source (irradiance, expressed in W/m²)
- The energy emitted from or reflected by a unit area of surface in a specified direction (irradiance, expressed in W/m²)

Although each of these methods of measurement can be used to effectively gauge illumination levels, most light source levels are expressed as radiance.

3.5.2 Noise Levels

Noise is any disturbing or unwanted sound and is among the most pervasive factors impacting the workplace, home, and recreational environments. Noise from occupational machinery, road traffic, construction equipment, manufacturing processes, and landscaping equipment, to name a few, is among the unwanted sounds that are routinely broadcast into the air.

The problem with noise is not only that it is unwanted, but it can also negatively impact human health and well-being. Health problems related to noise include noise induced hearing loss (NIHL), stress, high blood pressure, sleep loss, distraction, and lost productivity, as well as a general reduction in the quality of life.

3.5.2.1 Measuring Noise Levels

Measuring noise levels in the occupational environment is one of the most important aspects of an effective workplace hearing conservation and noise control program (Maxpro Maintenance, 2009). It helps identify work locations with excessive noise levels, employees who may be affected, and where additional noise mitigation methods need to be implemented.

Weighted sound level meters (SLMs) are used to measure noise (Figure 3.7). The weighted scales filter out the sound energy in the lowest and highest frequencies because human sensitivity in these bands is lower. The weighted decibel (db) scale

FIGURE 3.7 Weighted sound levels meters.

is most often used because studies have shown that it is a reliable determinant of noise level. Noise level considers the volume and frequency of the audible stimuli. The equivalent level of sustained noise (continuous sound level) is the summated frequency level.

$$L_{eq} = L_{50} + 0.43(L_1 - L_{50})$$

where
  L_{50} is the average noise level
  L_1 is the peak noise level

The equivalent level of sustained noise (L_{eq}) expresses the average level of sound energy during a given period of time. The summated frequency level is measured with a sound level indicator and a frequency counter, operating over a given time period. Typically, this time period is equal to a full 8 h work shift. A number of devices are capable of calculating this noise equivalent (L_{eq}), including noise dosimeters, SLMs, and integrated sound level meters (ISLMs). The type of sounds of interest and the source of the noise should be considered when determining how to measure the noise. Useful guidelines for instrument selection can be found in Table 3.9 from the Canadian Centre for Occupational Health Safety (2009).

3.5.2.1.1 Sound Level Meter

This device uses a microphone to pick up small, instantaneous variations in air pressure, and converts them into electrical signals. These signals are further processed until a decibel sound level is calculated. This dB level is captured and provided on the readout display. Calibration is required prior to each use and the SLM is commonly used to conduct noise surveys.

3.5.2.1.2 Integrated Sound Level Meter

The ISLM is very similar to the dosimeter (see the following text), except the device is handheld rather than worn. The ISLM is designed to perform calculations at a given static location, yielding a single noise reading despite any continuous fluctuations in the environment.

3.5.2.1.3 Noise Survey

A noise survey uses sample noise measurements from various locations to identify areas, machines, and equipment that generate harmful noise levels. This is usually done with a SLM. A sketch showing the locations of workers and noisy machines is drawn. Noise level measurements are taken at a representative number of positions around the area and are marked on the sketch. More measurements will yield an increasingly accurate survey and an accompanying noise map.

A noise map can be produced by drawing lines on the sketch between points of equal sound level. Noise survey maps, like that in Figure 3.8, provide very useful information by clearly identifying areas where there are noise hazards. The map is created by connecting points of equal sound level.

TABLE 3.9
Guidelines for Instrument Selection

Type of Measurement	Appropriate Instruments (in Order of Preference)	Result	Comments
Personal noise exposure	1. Dosimeter	Dose or equivalent sound level	Most accurate for personal noise exposures
	2. ISLM	Equivalent sound level	If the worker is mobile, it may be difficult to determine a personal exposure, unless work can be easily divided into defined activities
	3. SLM	dB(A)	If noise levels vary considerably, it is difficult to determine average exposure. Only useful when work can be easily divided into defined activities and noise levels are relatively stable all the time
Noise levels generated by a particular source	1. SLM	dB(A)	Measurement should be taken 1–3 m from source (not directly at the source)
	2. ISLM	Equivalent sound level dB(A)	Particularly useful if noise is highly variable; it can measure equivalent sound level over a short period of time (1 min)
Noise survey	1. SLM	dB(A)	To produce noise map of an area; take measurements on a grid pattern
	2. ISLM	Equivalent sound level dB(A)	For highly variable noise
Impulse noise	1. Impulse SLM	Peak pressure dB(A)	To measure the peak of each impulse

Source: Canadian Centre for Occupational Health Safety, 2009, Retrieved September 9, 2009, from http://www.ccohs.ca/

A standard SLM takes only instantaneous noise measurements. This is sufficient in workplaces with continuous noise levels. But in workplaces with impulse, intermittent, or variable noise levels, the SLM makes it difficult to determine a person's average exposure to noise over a work shift. One solution in such workplaces is a noise dosimeter.

3.5.2.1.4 Noise Dosimeter

A noise dosimeter is a portable sound measuring device for employees who are fairly mobile. A noise dosimeter is usually held in place near the hip, with a miniature microphone attached to the collar, as shown in Figure 3.9. This type of noise measurement is especially useful when the worker is changing locations frequently.

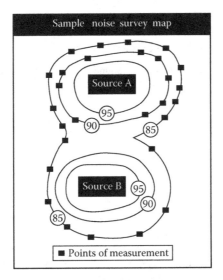

FIGURE 3.8   Sample noise survey map. (From Canadian Centre for Occupational Health Safety, 2009, retrieved September 9, 2009, from http://www.ccohs.ca/)

FIGURE 3.9   Noise dosimeter. (From Work Safe Saskatchewan, 2009, retrieved September 9, 2009, from http://www.worksafesask.ca/files/ont_wsib/certmanual/battery.jpg)

A noise dosimeter requires the following setting (Canadian Occupational Health Association, 2010):

- *Criterion level:* exposure limit for 8 h/day 5 days/week.
- Criterion level is 90 dB(A) for many jurisdictions, 85 dB(A) for some, and 87 dB(A) for Canadian federal jurisdictions.
- *Exchange rate:* 3 or 5 dB as specified in the noise regulation.
- *Threshold:* noise level limit below which the dosimeter does not accumulate noise dose data.

TABLE 3.10

Permissible Noise Exposures

Duration per Day, Hours	Sound Level dBA Slow Response
8	90
6	92
4	95
3	97
2	100
1½	102
1	105
½	110
1/4 or less	115

Source: Taken from the U.S. Department of Labor website under OSHA Section 1910.95(b)(1), Occupational Noise Exposure, http://www.osha.gov/pls/oshaweb/owadisp. show_document?p_table=STANDARDS&p_ id=9735&P_TEXT_VERSION=FALSE

Wearing the dosimeter over a complete work shift gives the average noise exposure or noise dose for that person. This is usually expressed as a percentage of the maximum permitted exposure.

3.5.2.2 Acceptable Noise Levels

The Occupational Safety and Health Administration (OSHA) publishes guidelines for acceptable noise levels in various settings, that is, noise levels that will not cause long-term harm to hearing, even for prolonged periods of time. Permissible noise exposure times and examples of different sound readings are found in Table 3.10.

It is often easier to understand a given sound level when it is compared to common activities. Table 3.11 provides examples of numbers, collected from a variety of sources that help one to understand the volume levels of various sources and how they affect our hearing (Table 3.11).

3.5.3 Thermal Conditions: Temperature and Humidity

Perceived thermal comfort is a function of environmental conditions and personal factors. The primary environmental factors are airflow (wind), air temperature, air humidity, and radiation from the sun or other radiant sources and nearby surfaces. The personal factors that affect comfort are clothing, level of physical activity, and level of acclimation or adaptation to the environment. Body heat balance then becomes a function of the personal, environmental, and administrative conditions. Table 3.12 provides a summary of the body heat balance factors and the heat balance equation for the body (Eastman Kodak, 1983).

The equation describing body heat balance is given below. Each term (column 1 in Table 3.12) is defined in column 2 of the table, and its major determinants are

TABLE 3.11
Decibel Comparison Chart

	Environmental Noise, dB
Weakest sound heard	0
Whisper quiet library	30
Normal conversation (3–5 ft)	60–70
Telephone dial tone	80
City traffic (inside car)	85
Train whistle at 500 ft, truck traffic	90
Subway train at 200 ft	95
Level at which sustained exposure may result in hearing loss	*90–95*
Power mower at 3 ft	107
Snowmobile, motorcycle	100
Power saw at 3 ft	110
Sandblasting, loud rock concert	115
Pain begins	*125*
Pneumatic riveter at 4 ft	125
Even short-term exposure can cause permanent damage—loudest recommended exposure with hearing protection	*140*
Jet engine at 100 ft, gun blast	140
Death of hearing tissue	180
Loudest sound possible	194

Source: Statistics for the decibel (loudness) comparison chart were taken from a study by Chasin, M., Centre for Human Performance & Health, Ontario, Canada, http://www.gcaudio.com/resources/howtos/loudness.html; University of North Florida, ADA Compliance, http://www.unf.edu/anf/adacompliance/Hearing_Campaign.aspx (accessed February, 2011).

identified in column 3. Conductive heat gain or loss (Co) has also been included at the end of the table, since this path of heat transfer is important when a hot or cold surface is directly contacted.

The heat balance equation for the body can be summarized as follows:

$$M \pm C \pm R - E = \pm S.$$

The level of humidity in an environment can significantly impact the comfort and perceived thermal load of operators. High humidity levels are often experienced in environments where water or excessive moisture is present as well as in outside working conditions. Table 3.13 provides a guideline for exposure time in certain humidity levels and ambient temperatures (Eastman Kodak, 1983).

3.5.3.1 Measurement of Thermal Load

Due to the variability involved in the perception of the thermal environment, there is no standard or regulation for workplace temperature in the United States. However, there are general guidelines that have been adopted by occupational health and

TABLE 3.12

Body Heat Balance

Term	Definition	Determinants
M	Metabolic heat gain	Physical workload, or the muscular work done minus the work efficiency
C	Convective heat gain or loss	Air velocity
		The difference between air temperature and a person's average skin temperature
R	Radiative heat gain or loss	The difference between a person's average skin temperature and the temperature of surfaces in the environment, measured with a globe thermometer or radiometer
		The amount of skin exposed to the solid surface
E	Evaporative heat loss	The difference between the water vapor pressure of a person's skin and the water vapor pressure, or relative humidity, of the environment; it is indirectly related to workload and the person's sweat rate
		Air velocity
S	Heat storage in, or loss from, the body	Balance of the aforementioned factors
		Rectal and skin temperatures
		Body weight
Co	Conductive heat gain or loss	The area of the conductive surface
		The difference between the person's skin temperature and the temperature of the surface contacted

Source: Developed from information in Kaman, 1975; Leithead and Lind, 1964; Eastman Kodak, 1986.

safety personnel. A commonly used reference is the threshold limit values (TLV) for heat stress, published by the American Conference of Governmental Industrial Hygienists (ACGIH). Specific to non-office environments, this publication is an excellent source of heat stress exposure literature. The units are expressed as wet bulb globe temperature (WBGT). The WBGT was developed in the late 1950s for the U.S. Marine Corps Recruit Depot on Parris Island in South Carolina (Australian Government website, 2010). Humidity in this region can be extreme and this measurement was developed to assess and minimize the impact on U.S. Marines in this environment.

The WBGT successfully combines temperature, humidity, and solar radiation to help employers estimate the environmental effects on workers. This composite measure resulted from studies following the 1950 fatal heat stroke outbreak at the U.S. Marine Corps Recruitment Depot on Parris Island in South Carolina. Over time, this index has been used in workplaces, as well as sporting events. More specifically, the WBGT is a culmination of the following:

• Dry bulb temperature
• Relative humidity

TABLE 3.13
Recommended Maximum Temperatures for Short-Duration Exposures to High-Heat Environments (up to 63°C or 146°F)

Exposure Time (min)	Workload[b]	Maximum Ambient Temperature, °C (°F)[a]		
		Relative Humidity 20%	Relative Humidity 50%	Relative Humidity 80%
5	L	63 (146)	56 (133)	56 (133)
	M	59 (138)	48 (118)	46 (115)
	H	57 (135)	46 (115)	42 (108)
15	L	53 (128)	45 (113)	40 (104)
	M	52 (126)	43 (110)	38 (100)
	H	51 (124)	41 (106)	36 (97)
30	L	52 (126)	44 (112)	39 (102)
	M	47(116)	38 (100)	34 (93)
	H	41 (106)	36 (97)	30 (86)
45	L	51 (124)	43 (110)	38 (100)
	M	41 (106)	36 (97)	31 (88)
	H	36 (97)	32 (90)	27 (81)

Source: Adapted from Rodgers and Corl, 1981, Eastman Kodak Company; based on information in Bell et al., 1971; Gagge, 1973; Hardy, 1970; Leithead and Lind, 1964; Pandolf and Goldman, 1978; Eastman Kodak, 1986.

Note: For 5 min exposure times in high air velocities (2 m/s, or 400 ft/min), the following maximum temperatures are recommended (L, light workload; M, moderate workload; H, heavy workload).

L	56 (133)	50 (122)	48 (118)
M	54 (129)	49 (120)	44 (111)
H	52 (126)	48 (118)	42 (103)

[a] These temperatures assume the following conditions: clothing insulation=0.6 clo; air velocity=0.1 m/s (20 ft/min); radiant heat=2°C (3.6°F), which is the difference between the globe and dry bulb.
[b] Workload abbreviations: L, light, up to 140 W (120 kcal/h); M, moderate, >140–230 W (>120–240 kcal/h); H, heavy, >230–350 W (>240–300 kcal/h).

- Mean radiant temperature
- Air velocity

The perceived temperature is calculated with consideration of these factors using two equations to calculate the WBGT:

$$WBGT = 0.7\,T_{nw} + 0.3\,T_g$$

where

T_{nw} is the natural wet bulb temperature obtained with wetted sensor exposed to natural air movement

T_g is the temperature of center of 6 in. diameter hollow copper sphere painted on the outside with black matte finish (globe thermometer)

For assessing WBGT when solar sources are present, the equation is modified to include this parameter:

$$WBGT = 0.7\,WB + 0.2\,GT + 0.1\,dB$$

where
  WB is the natural wet bulb temperature (temperature of sensor in a wet wick exposed to air current)
  GT is the globe temperature (globe temperature of a black sphere)
  dB is the dry bulb temperature (dry bulb temperature measured while shielded from radiation)
  Although several approaches have been developed, the standard methods and procedures to measure the factors in the WBGT calculation are fairly consistent. A discussion of the equipment is presented in the following section.

3.5.3.2 WBGT Equipment

WBGT instruments are available commercially and require regular maintenance if they are to consistently provide accurate measurements. The WBGT is measured by a three-temperature element device (Figure 3.10):

FIGURE 3.10   WBGT measurement equipment.

- The natural wet bulb temperature (WB) signifies the combined effects of wind, humidity, and radiation. The device consists of an unshielded thermometer with its bulb wrapped in a cotton wick. The wick is fed with a constant reservoir of distilled water. The continuous evaporative cooling of the wick is representative of sweat evaporation.
- The black globe temperature (GT) represents the combined effects of radiation and wind. The device is usually a 150 mm (6 in.) black globe with a centrally located thermometer.
- The (shade) air temperature (dB) is the standard forecast temperature. This consists of a radiation-shielded thermometer—usually contained in a weather screen.

The WBGT is not without criticism as many believe that the occupational environment has factors that impact thermal comfort that are not considered or sufficiently accounted for in this measurement. To address some of these issues, WBGT allows for correction factors with respect to clothing. A summary of correction factors for specific clothing types can be seen in Table 3.14. This table provides the clo, a value or insulation rating associated with the clothing type as described.

3.5.3.3 Other Thermal Indices
The effective temperature (ET) index is a culmination of temperature, humidity, and air velocity. This measurement is regularly used in mines and other places with high humidity and low radiant heat.

The heat stress index (HSI) considers the rate of work and environmental factors but is still unsatisfactory in measuring an individual's heat stress and can be difficult to implement.

Operative temperature takes into account the combined effects of radiation and convection but not humidity and air flow.

Oxford index is a weighting of the wet bulb and dry bulb temperatures (Parsons, 2003). This index is useful to account for the impact of variations on the worker and

---

TABLE 3.14
WBGT Correction Factors (°C)

Clothing Type	Clo[a] Value	WBGT Correction
Summer lightweight working clothing	0.6	0
Cotton coveralls	1.0	−2
Winter work clothing	1.4	−4
Water barrier, permeable	1.2	−6

Source: ACGIH, 1992.
[a] Clo: Insulation value of clothing. One clo = 5.55 kcal/m2/h of heat exchange by radiation and convection for each °C difference in temperature between the skin and the adjusted dry bulb temperature.

occupational factors related to thermal issues. Table 3.15 provides a summary of thermal strain indices (Bentel and Santee, 1997).

Each of these indices can be utilized to assess key aspects of the thermal environment on the worker. Careful consideration of the thermal environment, as well as economic factors, will impact the selection of the most appropriate approach.

3.5.3.4   Controlling the Thermal Conditions

There are five generally used engineering controls that reduce heat stress including ventilation, air cooling, fans, shielding, and insulation. Additional approaches can be used to reduce physical demands, which indirectly affect heat stress, provided there is a reduction in metabolic effort. Also, as previously discussed, the occupational environment should permit time for personal conditions such as acclimatization and adaptation to occur. Finally, fluid replacement and task design are controls that should be implemented as well.

3.5.3.4.1   Engineering Controls

- *General ventilation*—This method of cooling is provided through a permanent system for large areas such as overhead fan systems, or portable ventilation devices for small areas. General ventilation works by taking cooler air, typically from outside the work area, and distributing it within the environment containing the warmer air.
- *Air treatment/air cooling*—This method removes heat from the air in order to reduce the temperature.
- *Air conditioning*—While this tends to be the most expensive method, air conditioning is very effective. To reduce costs, chillers are often used as a substitute in cool or dry climates to cool the air. This system circulates cold water through coils in the air-conditioning system and distributes this cooled air into the environment to reduce the temperature indoors.
- *Local air cooling*—This method tends to be more effective at cooling smaller, targeted areas. This approach is generally less expensive to set up and can be quickly accomplished with a portable blower and air chiller.
- *Shielding or insulation*—Radiant heat can be reduced through shielding, or interrupting the path between the worker and the heat source. Radiant heat can be reduced through surface modification (i.e., a flat black surface will absorb more heat than a smooth, polished one), insulation, and shielding.

It is also useful to combine these techniques to amplify the reduction efforts to minimize the impact of the heat source.

3.5.3.4.2   Administrative Controls and Work Practices

Management participation and training are vital to implementing new work practices and promoting a healthy workplace. The following components are listed by NIOSH (1986) as being vital in creating a good heat stress training program:

TABLE 3.15
Thermal Strain Indices

Index	Source(s)	Inputs	Comments
		Indices for Heat Exposure	
Wet bulb globe temperature (WBGT)	ISO 7243 (1989); NIOSH (1986); Yaglou and Minard (1957); Botsford (1971)	T_a, T_{nwb}, T_g botsball	WBGT requires simple input and calculations. WBGT is not recommended for conditions of high humidity. The botsball is an instrument consisting of a fabric-covered 60 mm black ball over a dial thermometer. WGT may be read off a scale or converted to WBGT if T_a is known
Wet globe temperature (WGT)			
Heat stress index (HSI)	Belding and Hatch (1955); ISO 7933 (1989)	T_a, T_{wb}, T_{bg} or T_r, V, M	HSI is the ratio of evaporative heat loss required to maintain a constant body temperature to the maximum amount of sweat that can be evaporated under the given climatic conditions. Required sweat rate index (S_r^*), a further development of HSI, is used in ISO 7933
Oxford index (WD)	Leithead and Lind (1964)	T_a, T_{wb}	WD originated from research on men performing rescue tasks in hot underground mines. WD can be used to determine tolerance times
		Indices for Cold Exposure	
Wind chill index (WC)	Siple and Passel (1945)	T_a, V	WCI is an index for the cooling rate of wind and air temperature on bare skin. There is no adjustment for solar radiation, clothing, or activity level
Required clothing insulation (IREQ)	Holmer (1984); ISO TR 11079 (1993)	T_a, T_r, V, $R.H.$, M	Clothing insulation required for survival ($IREQ_{min}$) with a t_{sk} of $30^\circ C$ and for maintaining thermal equilibrium ($IREQ_{neutral}$) are calculated
		Equivalent Temperature Indices	
Effective temperature (ET)	Houghten and Yaglou (1923)	T_a, T_{wb}, V	ET relates actual conditions to an equivalent, calm, saturated environment. ET overemphasizes effects of humidity in cool and neutral conditions and underemphasizes its effects in warm conditions

(continued)

TABLE 3.15 (continued)
Thermal Strain Indices

Index	Source(s)	Inputs	Comments
"New" effective temperature (ET*)	ASHRAE (1993); Gagge et al. (1971); Gonzalez et al. (1974)	T_a, T_r, V, P_a, i_m, w, M	ET* was developed to replace ET. It includes skin-wettedness (w) and water vapor pressure (P_a) parameters in calculating the temperature of an environment at 50% R.H. that results in equivalent total heat loss from the skin as in the actual environment
Operative temperature (T_o)	Winslow et al. (1937)	T_a, T_{bg} or T_r, V	T_o combines dry heat exchange parameters. There is no adjustment for work rate or effects of humidity on evaporative cooling. Formulas for approximating T_o are provided in ISO 7730 (1993)

Source: Gavriel, S. (Ed.): *Handbook of Human Factors and Ergonomics*, 1997. Copyright Wiley-VCH Verlag GmbH & Co. KGaA.

- Knowledge of the hazards of heat stress
- Recognition of predisposing factors, danger signs, and symptoms
- Training of employees regarding their responsibilities in avoiding heat stress
- Awareness of first-aid procedures for, and the potential health effects of heat stroke
- Dangers of using drugs, including therapeutic ones, and alcohol in hot work environments
- Use of protective clothing and equipment
- Purpose and coverage of environmental and medical surveillance programs and the advantages of worker participation

It is also important to observe NIOSH recommended WBGT limits in environments prone to produce heat stress. Table 3.16 provides exposure guidelines from NIOSH for acclimated workers (NIOSH, 1986).

Additional administrative controls include scheduling outdoor jobs for the cooler part of the day, adding breaks at appropriate intervals, job sharing, and fluid replacement options throughout task performance. This includes worker monitoring activities to further reduce the likelihood for heat stress. Factors affecting the likelihood of heat stress are varied and in some cases cannot be administratively controlled and are summarized in see Table 3.17 (NIOSH, 1986).

The primary issues to consider in the development of a worker monitoring program to manage thermal stress include the following:

1. Identifying tasks and workers with an increased risk of heat stress
2. Establishment of a plan and process to do heat stress monitoring regularly
3. Define process for monitoring
4. Monitoring may include any combination of measures including
   a. Measuring heart rate

TABLE 3.16

NIOSH Recommended WBGT Limits, in °C, for Heat Stress Exposure for Acclimatized Workers

Hourly Work/Rest Cycle	Workload		
	Light (<230W)	Moderate (230–350W)	Heavy (>350W)
Continuous work	<30.0	<26.7	<25.0
75% work/25% rest	30.6	27.8	25.6
50% work/50% rest	31.7	29.4	27.8
25% work/75% rest	32.2	31.1	30.0
Ceiling limit	38.9	36.7	35.0

Source: Gavriel, S. (Ed.): *Handbook of Human Factors and Ergonomics*, 1997. Copyright Wiley-VCH Verlag GmbH & Co. KGaA; NIOSH, 1986.

Note: Limits are for a "standard" worker weighing 70 kg with a 1.8 m² body surface area.

TABLE 3.17

Factors Affecting the Occurrence of Heat Stress

Factor	Effect
Hydration state	Hypohydration results in lower sweat production and an increase in core temperature
Acclimatization	Repeated heat exposure leads to earlier onset of sweating, a higher sustained sweat rate, and lower core temperature and heart rate
Age	Sweating mechanism and circulatory system become less responsive with age, and there is high level of skin blood flow, possibly due to impaired thermoregulatory mechanism
Physical fitness	Exercise that increases maximal aerobic capacity improves thermoregulatory responses in the heat
Subcutaneous fat	Subcutaneous fat provides an insulative barrier, reducing transfer of heat from muscles to skin
Gender	Although studies indicate that sweating and vasodilation occur at higher core temperatures in women than men, when controlled for fitness and menstrual phase, gender differences in the follicular phase are questionable. Women in the luteal phase have significantly higher core temperatures, which may impact thermoregulatory responses
Body size	Leaner individuals are at an advantage because they have a larger ratio of surface area to body mass and, thus, greater capacity to dissipate heat
Diet	Regular consumption of a balanced diet serves to replace salt and other electrolytes lost in sweat, maintaining sweating efficiency
Previous heat illness	Previous occurrence of heat stroke increases susceptibility to subsequent heat illness
Drugs and alcohol	Use interferes with the functioning of the central and peripheral nervous system, negatively affecting heat tolerance

Source: Gavriel, S. (Ed.): *Handbook of Human Factors and Ergonomics*, 1997. Copyright Wiley-VCH Verlag GmbH & Co. KGaA.

    i. Working heart rate—obtained by counting the radial pulse for 30 s immediately after work has ceased.

    ii. Recovery heart rate—obtained by comparing the initial 30 s pulse to a second pulse taken 2.5 min into the rest period.

  b. Oral temperature—measured with a clinical thermometer after work but before the employee drinks water. If the oral temperature taken under the tongue exceeds 37.6° C, shorten the next work cycle by one-third.

  c. Body water loss—can be measured by weighing workers prior to work, at different intervals during the work shift, and at the conclusion of the work shift.

3.6   VIBRATION AND THE HUMAN BODY

The International Standard (ISO 2631) for evaluation of human exposure to whole-body vibration gives guidelines for how vibration should be measured and assessed.

The ISO 2631 standard entitled "Mechanical vibration and shock—Evaluation of human exposure to whole-body vibration" consists of four parts (2631-1, 2631-2, 2631-4, and 2631-5) (Nakashima, 2004). The impact of vibration on the human body is considered for varied ranges and can produce numerous outcomes on the physical and sensory capabilities. The three ranges of frequency–classification in the ergonomic assessment of vibration are

- Low frequencies (0 to 2 Hz)
- Middle frequencies (2 to 15–20 Hz)
- High frequencies (greater than 20 Hz)

Historically, the occupational safety and ergonomic communities were much more concerned with high-frequency vibration; however, recent years have revealed a need to be concerned about all types of vibration. It is well documented that the low and middle frequencies can cause at least comparable damage, and in some cases *more* cumulative trauma to the human body than vibrations at the higher frequencies.

The human body can amplify the vibration (in these lower frequencies) that exists from an outside source. We know, for instance, that the musculoskeletal system (muscles, tendons, and bones) can "be a path" for vibration and actually amplify the vibration as it moves through the body. An example of low-frequency vibration includes the use of handheld power tools that transmit the vibration from the hand, through the arm, and to the upper body.

Human response to vibration depends on several factors, including the frequency, amplitude, direction, point of application, time of exposure, clothing and equipment, body size, body posture, body tension, and composition. A complete assessment of exposure to vibration requires the measurement of acceleration in well-defined directions, frequencies, and duration of exposure (WISHA Hand-Arm Vibration Analysis, 2010).

Vibration measurement instruments can be used to measure various types of vibration including hand-arm, foot-leg, or full body impact. A typical vibration measurement system includes an accelerometer, which is a device to sense the vibration, a recorder, a frequency analyzer, a frequency-weighting network, and a display, such as a meter, printer, or recorder. The accelerometer produces an electrical signal in response to the vibration. The size of this signal is proportional to the acceleration applied to it. The frequency analyzer determines the distribution of acceleration in different frequency bands. The frequency-weighting network mimics the human sensitivity to vibration at different frequencies. The use of weighting networks gives a single number as a measure of vibration exposure (i.e., units of vibration) and is expressed in meters per second square (m/s^2).

The cumulative effects on the body are seen in musculoskeletal disorders, vestibular problems, and other disorders. Some of the more common disorders are white finger disease (Raynaud's syndrome), hand-arm vibration syndrome, carpal tunnel syndrome, tendonitis, and various bone and joint disorders. Long-term exposure to vibration can lead to nausea, impaired vision, hyperventilation, high blood pressure, impaired cardiac rhythm, and premature fatigue.

The methods to reduce the exposure or result of vibration on the user should be aggressive, as the conditions that result can be debilitating and long term. Techniques that can be used to mitigate the effects of vibration include source control, path control, and receiver control (Wasserman and Wilden, 2006).

- *Source control* is considered a long-term engineering approach to address the source (tool, equipment, or vehicle) that is emitting the vibration. This should ideally be accomplished at the design stage; however, it is often required in redesign of existing products or task environments. In addition to design changes to reduce vibration intensity, source control can be accomplished with wraps on the tool, regulation of tool speed, or balancing and process redesign.
- *Path control* is accomplished by altering the amount of exposure to the vibration. To reduce the amount of vibration, administrative controls can be applied to alter the task performance process. This can be accomplished by rotating personnel, job sharing, and providing frequent rest breaks throughout the tasks.
- *Receiver control* is designed to reduce the vibration experienced by the operator. To reduce vibration, the receiver control approach isolates tools to reduce the level of vibration that is received or adapts posture to limit exposure. Other means to control vibration received by the operator include modifying the task to allow a reduction in force exertion (i.e., grip, grasp, or push forces) or reducing contact area for task performance. Personal protective equipment, such as padded gloves, can also reduce the impact of the vibration.

3.7 EXPOSURE TO CHEMICALS, RADIATION, AND OTHER SUBSTANCES

Many jobs require employees to be exposed to chemical, biological, radiation, and other types of hazards. Examples of tasks with these risk factors include waste disposal, recycling professions, health-care professions, and first responders. Exposure to hazardous substances can have serious effects on workers' health. Some substances, such as asbestos (which can cause lung cancer and mesothelioma), are now banned or subject to strict control worldwide. Yet many substances that are still widely used can also cause serious health problems, if the risks associated with them are not managed. The impact of these substances can have a wide range of health effects including (OSHA Europa, 2010)

- Acute effects, for example,
  - Poisoning
  - Suffocation
  - Explosion
  - Fire
- Long-term effects, for example,
  - Respiratory diseases (i.e., reactions in the airways and lungs) such as asthma, rhinitis, asbestosis, and silicosis
  - Occupational cancers (i.e., leukemia, lung cancer, mesothelioma, cancer of the nasal cavity)

- Health effects that can be both acute and long-term, for example,
  - Skin diseases
  - Reproductive problems and birth defects
  - Allergies
- Accumulation in tissues
  - Some substances can accumulate in the body (e.g., heavy metals such as lead and mercury or organic solvents) gradually poisoning the system
  - Some substances can produce long-term health and reproductive issues
- Penetration through the skin and dermal conditions

Workers regularly exposed to liquids, including water, which can break down the skin's natural defense barrier, are most at risk of developing skin problems.

Exposure to extreme temperature, solar radiation, and biological risks also contribute to exposure-related problems. Heavy physical work can also enhance the uptake of dangerous substances.

3.8 SUMMARY

The human body is a complicated and amazing collection of systems and sensory capabilities that permit an infinite number of movements, activities, and functions. The senses of the body are an inherent safety mechanism, which respond to external stimuli and are the gateway to the human. The senses interpret external signals and allow the internal systems to respond. Hazardous conditions may be immediately sensed and workers and management must pay attention to these signals to eliminate acute hazards. Some hazards pose risks that may not be immediately detected. These types of hazards can cause long-term disability and health concerns. These dangers are even more critical for engineers and management to consider in workplace design, as workers may not realize in the short term the long-lasting consequences.

These senses of the body are the pathway for human perception of the outside world.

The body responds to stress from internal and external stimuli. It is important to reduce external stimuli so bodily systems are not overly stressed. This is called "engineering" the problem out of the task design. It is equally important for management and employees to recognize hazards in the workplace that may stress the body beyond acceptable limits and respond to the signals and signs a body exhibits when it is being unduly stressed. Mitigation of risks should always be a priority.

Case Study

Ergoweb® Case Study—Skin Stapler Assembly and Welding Operation
By Arthur R. Longmate and Timothy J. Hayes

Task Prior to Abatement (Description)

This task was performed by a four-man crew (two assemblers, one welder, and one cleaner/packer). With the tote upright on the table, the assembler would reach into the pan each time and get a single component (total five components) for each assembly. Workers needed to reach near the bottom of the tote when the pans were less than half-full. The welding operation involved welding the instrument in an ultrasonic welder and then firing it five times to test staple formation and staple feed in the magazine.

Task Prior to Abatement (Method Which Verified Hazard)

The welding operation steps were:

1. Get one instrument from tray with the left hand
2. Position and insert into welder nest using the same hand
3. Close manual clamp on welder nest with left hand
4. Push and hold welder activation buttons to cycle welder
5. Get instrument from weld nest using right hand
6. Fire instrument once by striking trigger forcefully with palm of left hand
7. Place instrument aside for final cleaning or destroy it if not acceptable
8. Record the defect type on sheet

If the welder station became backed up, one of the assemblers would swing over and assist the welder by performing the test-firing function.

Task Prior to Abatement (Method Which Identified Hazard)

The ergonomic-related medical incidence rate was extremely high in this department. These incidences include various forms of tendinitis and hand/wrist related disorders.

Many employees were placed on medical restrictions.

Ergonomic Risk Factor (Force)

High mechanical force concentrations to the hands and fingers and high hand force is required to dig the parts out.

Ergonomic Risk Factor (Posture)

Longer reach than necessary and difficulty in grasping is required to grasp the parts.

Ergonomic Risk Factor (Repetition)

High repetitive wrist flexion and ulnar deviation is required in order to strike the trigger forcefully with the palm of the left hand five times to fire each instrument for approximately 4000 instruments per day.

Ergonomic Solution (Administrative Controls)

A structured job rotation sequence was established to reduce the exposure to the high repetitive and forceful tasks.

Ergonomic Solution (Engineering Controls)

Assembly stations with adjustable V-stands were provided to tilt up tote pans to the most accessible angle without allowing parts to spill out onto the table.

New, adjustable ergonomic chairs were purchased and footrests were provided for shorter employees to reduce the risk of posture related injuries and increase comfort.

A presence sensing activation button system was provided and an adjustable angle mounting bracket was developed to attach the activation buttons to the side of the welder in order to reduce the high repetition and forceful task of activating buttons using the thumbs (reducing the thumb tendinitis).

A pneumatic clamp was used to automatically clamp the instrument in order to eliminate repetitive striking of a manual De-sta-co clamp to retain the instrument for welding.

Ergonomic Solution (Benefits)

All workers that perform this task now have reduced exposure to various forms of tendinitis and other hand/wrist related disorders.

- Productivity has increased up to 10%–12%.
- Employee response to the modifications has been extremely positive and they can perform the task with less risk of injury.

Ergonomic Solution (Method Which Verified Effectiveness)

There has been a 10%–12% increase in productivity.

New medical problems have greatly diminished.

Comments

Providing a conveyor to automatically transport trays of assembled instruments between the assembly and welding workstations is under consideration, although from an ergonomic view, the positive influence of a conveyor is not as clear because of the low weight (about 5 lb) of the full trays of instruments. Due to the relatively high cost of the conveyor (about $8000 per line) and due to the questionable ergonomic impact, justification is in question.

Source: Longmate, A.R. and Hayes, T.J., *Ind. Manage.*, 32(2), 27, March/April 1990.

EXERCISES

3.1 Explain an occupational environment where each of the human senses is utilized for task performance.

3.2 Explain how to minimize overloading of the visual system in a visual inspection task in a manufacturing facility.

3.3 Discuss the types of vibrations impacting workers and the sources of vibration. How can these levels of vibration be determined?

3.4 Why do many older people need higher illumination to see objects clearly?

3.5 Explain the impact of new recreational technology on the risk for NIHL.

3.6 Identify a national or international standard (or guideline) for each occupational design in vision, audition, and vibration.

3.7 What approach should be used to estimate thermal loading in an outside task?

3.8 Discuss risk factors associated with the use of the gestation sense in task performance.

3.9 Explain the relationship between the olfactory and gestation senses.

3.10 Describe a task that relies largely on the sense of tactation.

REFERENCES

Australian Government Website. (2010). http://www.bom.gov.au/info/thermal_stress/

Caceci, T. (n.d.). Connective tissues III: Cartilage types. Retrieved September 9, 2009, from http://education.vetmed.vt.edu/curriculum/vm8054/labs/Lab7/IMAGES/CTL23.JPG

Canadian Centre for Occupational Health Safety. (2009). Retrieved September 9, 2009, from http://www.ccohs.ca/

Canadian Occupational Health Association. 2010 http://www.ccohs.ca/oshanswers/phys_agents/noise_measurement.html

Chasin, M., Centre for Human Performance & Health, Ontario, Canada, http://www.gcaudio.com/resources/howtos/loudness.html

Correa, J. (2005). The dog's sense of smell, Alabama Cooperative Extension System, PubID: UNP—Online Publication. http://www.aces.edu/pubs/docs/u/UNP-0066/ (accessed September, 2011).

Fanger, P.O. (1988). Introduction of the olf and the decipol units to quantify air pollution perceived by humans indoors and outdoors. *Energ. Buildings*, 12(1), 1–6.

Gavriel, S. (Ed.) (1997). *Handbook of Human Factors and Ergonomics*, Wiley Interscience: New York.

Gottfried, J., Smith, A., Rugg, M., and Dolan, R. (2004, May 27). Remembrance of odors past. *Neuron*, 42(4), 687–695.

Hitti, M. (2007, March 9). Smelling rose scent may strengthen memory during sleep. WebMD. Retrieved January 23, 2001, from http://www.webmd.com/brain/news/20070309/scent-strategy-improving-memory

http://www.springerlink.com.proxy.library.emory.edu/content/0301-5548/ *European Journal of Applied Physiology and Occupational Physiology* http://www.springerlink.com.proxy.library.emory.edu/content/0301-5548/78/2/ *Volume 78, Number 2,* 93-98, DOI: 10.1007/s004210050392. http://www.springerlink.com.proxy.library.emory.edu/content/0301-5548/

http://www.sciencedirect.com.proxy.library.emory.edu/science/journal/00016918, *Acta Psychologica,* (accessed April 12, 2010).

http://www.sciencedirect.com.proxy.library.emory.edu/science?_ob=PublicationURL&_tockey=%23TOC%235795%232001%23998929998%23245097%23FLA%23&_cdi=5795&_pubType=J&view=c&_auth=y&_acct=C000034138&_version=1&_urlVersion=0&_userid=655046&md5=b4d6f33b77b88aa8022c9a6d6ece3590, 107(1–3), April 2001, 9–42, (accessed March 28, 2011).

http://dx.doi.org.proxy.library.emory.edu/10.1016/S0001-6918%2801%2900018-X, doi:10.1016/S0001-6918(01)00018-X , (accessed March 28, 2011).

Iowa State University. (2004, July 12). The science of smell part 2: Odor chemistry. Retrieved January 23, 2011, from http://www.extension.iastate.edu/Publications/PM1963B.pdf

Kaufman, J.E. and Christensen, J.F. (1972). *IES Lighting Handbook: The Standard Lighting Guide*, Illuminating Engineering Society (U.S.): New York.

Kodak, E. (1983). *Ergonomic Design for People at Work*, Eastman Kodak: Rochester, NY.

Konz, S.A. (1997). Toxicolgical and thermal comfort. In: G. Salvendy (ed.), *Handbook of Human Factors*, Wiley: New York.

Kroemer, K.H.E. and Grandjean, E. (1997). *Fitting the Task to the Human: A Textbook of Occupational Ergonomics*, CRC Press: Boca Raton, FL.

Lariviere, C., Gravel, D., Arsenault, A.B., Gagnon, D., and Loisel, P. (2003). Muscle recovery from a short fatigue test and consequence on the reliability of EMG indices of fatigue. *Eur. J. Appl. Physiol.*, 89(2), 171–176. Published online. http://www.ncbi.nlm.nih.gov/pubmed/12665981 (accessed September, 2011).

Longmate, A.R. and Hayes, T.J. (1990, March/April). Making a difference at Johnson & Johnson: Some ergonomic intervention case studies. *Ind. Manage.*, 32(2), 27.

Mann, M. Hear, boy: Why humans can't hear everything animals can, Divine Caroline: Life in Your Own Words, http://www.divinecaroline.com/22178/102125-hear-boy-humans-can-t-hear/3#ixzz1DxBiOBCI

Maxpro Maintenance. 2009. Monarch 322 kit. Retrieved September 9, 2009, from http://www.maxpromaintenance.com/images/Monarch322Kit.jpg

Morgan, C.T., Cook, J.S., Chapanis, A., and Lund, M.W. (1963). *Human Engineering Guide to Equipment Design*, McGraw-Hill: New York.

Myers, D.G. (2004). *Exploring Psychology*, Worth Pub: New York.

Nakashima, A. (2004, July). The effect of vibration on human performance and health: A review of recent literature. DEFENCE R&D CANADA, Technical Report, DRDC Toronto TR 2004-089.

National Institute for Occupational Safety and Health (NIOSH). (1986).

OSHA Europa. (2010). http://osha.europa.eu/en/topics/ds/health_effects

Pancaldi, G. (2003). *Volta: Science and Culture in the Age of Enlightenment*, Princeton University Press: Princeton, NJ.

Parsons, K. (2003). *Human Thermal Environments: The Effects of Hot, Moderate, and Cold Environments on Human Health, Comfort, and Performance*, CRC Press: Boca Raton, FL.

Raudenbush, B. (2001). Enhancing athletic performance through the administration of peppermint odor. *J. Sport Exerc. Psychol.*, 23, 156–160.

Sense, *New World Encyclopedia*. (2008, August 29). Retrieved 22:06, January 23, 2011 from http://www.newworldencyclopedia.org/entry/Sense?oldid=795149

Smith, D.V. and Margolskee, R.F. (2001, March). Making sense of taste, *Scientific American*, pp. 32–39.

Teamwork Photo & Digital. (2009). Sekonic. Retrieved September 9, 2009, from http://www.teamworkphoto.com/images/sekonic/758d.jpg

University of North Florida, ADA Compliance, http://www.unf.edu/anf/adacompliance/Hearing_Campaign.aspx

Vogel, I., Brug, J., Hosli, E.J., van der Ploeg, C.P.B., and Raat, H. (2008). MP3 players and hearing loss: Adolescents' perceptions of loud music and hearing conservation. *J. Pediatr.* 152(3), 400,404.e1

Wasserman, D. and Wilder, D. (2006). Vibrometry, *Occupational Ergonomics Handbook*, 2nd edn. (Eds. W. Karwowski and W. Marras), Chapter 33, Taylor & Francis: New York.

WISHA Hand-Arm Vibration Analysis. (2010). State of Washington Department of Labor and Industries Ergonomics Rule. http://personal.health.usf.edu/tbernard/hollowhills/WISHA_HAV. pdf (accessed November, 2010).

Woodson, W.E., Tillman, B., and Tillman, P. (1992). *Human Factors Design Handbook*, 2nd edn., McGraw-Hill, Inc.: New York.

4 Muscular Work and Nervous Control of Movements

4.1 LEARNING GOALS

This chapter will provide an explanation of the functionality, relationship, and elements of the integrated roles of the muscular system and nervous system. The execution and control of movements as a function of these systems will also be provided to gain an in-depth understanding of these processes.

4.2 KEY TOPICS

- Muscular work
- Muscular contractile system
- Mechanism of contraction
- Method to stimulate and control the mechanism of contraction
- Energy to drive mechanism of contraction
- Innervation of the muscular system
- Reflexes
- Energy transformation process for muscle activity
- Types of muscular work
- Muscular fatigue
- Types of muscle contraction
- Measurement of muscular strength

4.3 INTRODUCTION AND BACKGROUND

The relationship between the musculoskeletal system and the nervous system is highly integrated in describing muscular work and movement. In light work or intense efforts, the communication and responses between the muscles and nerves enable the human system to perform tasks ranging from simple to extremely complex, requiring the coordinated action of both the nervous and muscle tissues. This was first shown by Galen Pergamum (129–199), a physician in ancient Greece, in experiments he conducted on animals. He demonstrated the importance of the central nervous system by cutting the spinal cord in varying positions, which resulted in the corresponding level of paralysis in the animal. Galen was also the first to identify muscles and show that muscles operate in combination as groups. This was further expanded on by von Haller (1708–1777), a Swiss physiologist who found that all nerves within the body eventually lead to the brain or spinal cord (Williams, 2009). This observation led him to conclude that the brain and spinal cord act as sensory perception and response action centers. He also demonstrated that muscles could be activated, that is, made to constrict at the introduction of a stimulus transmitted by a nerve. Later, in 1780, Luigi Galvani (1737–1798) noticed while dissecting the legs of a frog that electrical stimuli caused the legs to twitch. This finding showed that the electrical impulses are involved in nerve and muscle simulation. Based on these principles, Galvani created a crude battery that combined metals and natural juices from a frog, which led his colleague, Alessandro Volta, to design the first electric battery (which, incidentally, did not require frog juice) (Pancaldi, 2003).

Today's neuromuscular research includes the controversial use of stem cell transplantation from umbilical cord blood and bone marrow to treat paralyzing spinal cord injuries and degenerative diseases. The moral and ethical issues surrounding stem cell research are aimed mostly at the closely related issues of cloning, reproductive techniques, and—at its most extreme interpretation—eugenics, which many fear could lead to the Hitler-era race purification of apocalyptic proportions. For this reason, scientific discovery in American stem cell research is often delayed. While many opponents of stem cell research exist within the United States, countries without such sociopolitical challenges have been able to make several medical breakthroughs with the use of stem cells. This has led to greater collaboration among American and non-American stem cell research institutions (Magnus and Cho, 2005).

4.4 MUSCULAR WORK

Muscle contraction is a function of a sophisticated muscle contractile process. The ability to voluntarily exert muscular force produces strength, which in turn produces work. The contractile process, energy exchange, and other musculoskeletal system actions integrate to accomplish the goals of the muscular work.

4.4.1 Muscular Contractile System

The contractile properties of muscles allow the tissue to execute a variety of mechanical activities with the bones and joints. A very unique characteristic of a muscle is its ability to shorten or contract to about half its normal resting length. Each muscle fiber contains proteins, including actin, myosin, tropomyosin B, and troponin. However, the two proteins that play the most important role in muscular contraction are actin and myosin. There are three aspects to the muscular contractile system and these include the following (Figure 4.1):

- Mechanism of contraction (sliding filament model)
- Method to stimulate and control the mechanism of contraction
- Energy to drive the mechanism of contraction

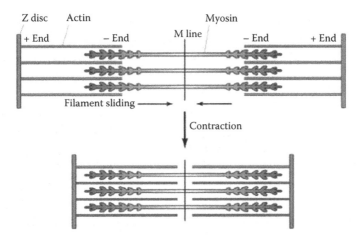

FIGURE 4.1   Sliding filament model.

4.4.1.1   Mechanism of Contraction (Sliding Filament Model)

For muscle fibers to shorten when they contract, they must generate a force that is greater than the opposing force that is acting to prevent movement of the muscle fiber insertion. The A bands within each muscle fiber are composed of thick filaments, while the I bands contain thin filaments. The movement of cross bridges that extend from the thick to the thin filaments produces the sliding of the filaments, resulting in muscle tension and shortening. The activity of the cross bridges is regulated by the availability of calcium, which is increased by electrical stimulation of the muscle fiber. Electrical stimulation produces contraction of the muscle through the binding of calcium to regulatory proteins within the thin filaments (Figure 4.2).

Before a muscular contraction takes place, the myosin and actin filaments are at an angle to each other. Each molecule of myosin in the thick filaments contains a globular subunit called the myosin head. The myosin heads have binding sites for the actin molecules in the thin filaments and adenosine triphosphate (ATP). The following steps describe the muscle contraction process:

- In order to initiate contraction, the myosin heads protrude from the thick filaments along its length and form a cross bridge with the actin.
- The myosin heads change configuration by rotating and pulling one filament against the next.
- The filaments "swivel" to a new angle, thus pulling the actin filaments past the thick or myosin filaments.
- The sarcomere shortens, which results in a muscular contraction.

As the muscle contracts, the Z lines come closer together, the width of the I bands and H zone decreases, and there is no change in the width of the A band. Conversely, when a muscle is stretched, the width of the I bands and H zone increases, but the width of the A band does not change (Huxley and Niedergerice, 1954).

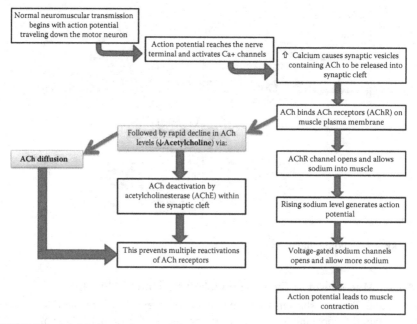

FIGURE 4.2   Model of muscle contraction.

4.4.1.2   Method to Stimulate and Control the Mechanism of Contraction

The number of actively contracting muscle fibers determines how much strength is developed during the period of contraction and regulates the speed of the muscular contraction or movement. A muscle produces its greatest active strength at the beginning of its contraction, when it is still near its relaxed length and as it shortens the ability to produce force declines. The number of nervous impulses determines the amount of muscle strength produced. During slow or maintained muscular contractions, muscle fibers are brought into active contraction in succession, allowing them to alternate. This gives the fibers some resting periods, which permits recuperation from fatigue.

A person's muscular strength depends on the cross section of his or her muscles. Each muscle fiber contracts with a specific force, thus men and women can become equally strong per cross section if both receive equal training. However, women, as a group, have narrower muscles and, therefore, on average are able to only about exert two-thirds the force of men.

4.4.1.3   Energy That Drives Contraction

Glucose and components of fat and protein are the indirect sources of energy required for the continuous replenishment of energy reserves in the form of ATP or other energy-rich phosphate compounds. Glucose is the main energy supply in intensive physical work, since it is immediately available and easily converted. Also, oxygen plays an important role in the production of energy in muscles.

As glucose passes out of the bloodstream and into the cells, it is converted into pyruvic acid. If oxygen is present, the pyruvic acid is broken down by oxidation, resulting in water and carbon dioxide. This process releases enough energy to restore large amounts of ATP. However, if there is a lack of oxygen, the breakdown of pyruvic acid will not take place. In this case, the pyruvic acid is converted into lactic acid, a form of metabolic waste product, which causes muscle fatigue. This process releases less energy for the restoration of energy-rich phosphate compounds and allows a higher muscular performance for a short time, under low oxygen conditions. An example would be when a person performs a heavy muscular task and becomes out of breath.

4.4.2 INNERVATION OF THE MUSCULAR SYSTEM

The contraction of skeletal muscle is controlled by the nervous system. As described in the previous section, the efferent and afferent nervous systems work in tandem to receive and send information between the peripheral nervous system and the central nervous system. The efferent (motor) nerves and the afferent (sensory) nerves support this communication process throughout the musculoskeletal system.

4.4.2.1 Efferent Nerves

Motor neurons leading to skeletal muscles have branching axons, each of which terminates in a neuromuscular junction with a single muscle fiber. Nerve impulses passing down a single motor neuron will trigger contraction in the muscle fibers, at which the branches of that neuron terminate. This group of muscle fibers, innervated by branches of the same efferent neuron axon, is called a motor unit. The motor unit is the functional unit of the muscle and varies in size based on the number of fibers innervated and the function of the muscle. A motor unit is small in muscles where precise control is important, such as in eye muscles. Motor units are larger in coarse acting muscles, such the biceps. The motor impulse leaps across from the nerve fiber to the muscle fiber at the motor end plates. The junction between the terminal of a motor neuron and a muscle fiber is called the neuromuscular junction. The motor end plates are the axonal effector endings, which initiate muscular contraction (Figure 4.3).

4.4.2.2 Sensory Nerves

Sensory nerves transmit impulses from the muscles into the central nervous system, either to the brain or to the spinal cord. Sensory impulses contain signals, which are utilized in the central nervous system to direct muscular work and to store as information. Muscle spindles, which run parallel to muscle fibers, are receptor organs that detect dynamic and static changes in muscle length and send sensory impulses to the spinal cord whenever a muscle changes in length. The Golgi tendon organ is another sensory receptor organ embedded in the tendons that. These organs monitor the tension that develops in a muscle and carry sensory impulses to the spinal cord whenever the tendon is under tension. When a sensory receptor is stimulated, the following process occurs:

- Sensory signals pass from the receptor along a sensory neuron to the spinal cord.

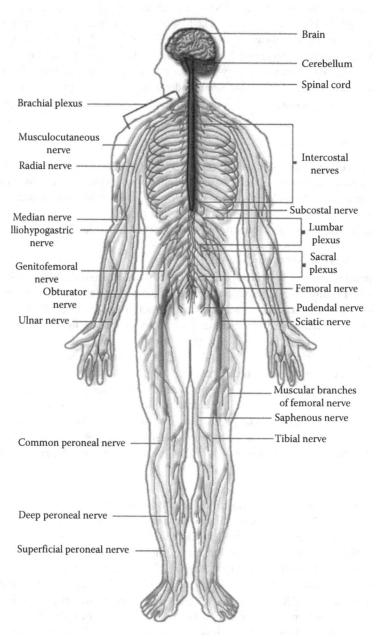

FIGURE 4.3 Efferent nerves.

- In the spinal cord, sensory impulses pass through the interneuron, which then makes contact with a motor neuron.
- Finally, the motor neuron carries efferent impulses to the muscle, which produces the response.

This pathway created from the afferent sensory nerves and efferent motor nerves in a muscle is called a reflex arc. The reflexes allow the muscle tension and muscle length to continually adapt and remain compatible; the muscle spindle and the Golgi tendon organ are the detectors in this regulatory system.

4.4.3 REFLEXES

Reflexes play crucial roles in certain muscular movements and activities. Reflexes are not consciously directed and, therefore, may be considered automatic in a physiological sense. A reflex comprises three parts:

- An impulse, which travels along a sensory nerve, carrying the information to the spinal cord or to the brain
- Interneurons, which transmit the impulse across to a motor nerve
- A final impulse along the motor nerve, which activates the appropriate muscle fibers producing the resulting movement

An example of a reflex is the blinking of the eyelids, when an unexpected object moves close to either eye. It is the unexpected movement in the visual field that provides the initial stimulus. The sensory impulse then travels to a particular center in the brain, where it acts as the interneuron and transmits the message to a motor nerve, which operates the muscles of the eyelids and causes the blink. The transmission time for neurons in some cases is greater than the recorded time for actual reflexes. Thus, it is suggested that the actions associated with a reflex may, in fact, be processed in the spinal cord rather than traveling the full distance to and from the brain (Hockenbury and Hockenbury, 2008).

Reflex blinking can be thought of as an automatic protective mechanism, which protects the eyes against any imminent damage. The human body uses thousands of similar reflexes, not just for protection, but also as part of normal control functions. A more complex reflex is the antagonistic control of a muscular movement between two muscles. For example, when the lower arm is bent, the bending muscles contract by stimulating motor nerves. In order for this muscle contraction to proceed smoothly, the opposing muscles behind the arm must be simultaneously relaxed by exactly the right amount. This co-contraction is an automatic reflex, producing a movement that is executed in a well-controlled manner.

4.4.4 ENERGY TRANSFORMATION PROCESS FOR MUSCLE ACTIVITY

In order for a person to do work, he or she has to make use of muscles. Muscles work only when there is energy to permit muscular contraction to take place, and the contractile process requires considerable energy. This energy, in the form of high-energy phosphate compounds, such as creatine phosphate (CP) and ATP, is transformed from nutrients by oxidative and non-oxidative processes known as metabolism.

During muscular work, chemical energy is transformed into mechanical energy. During this transformation the mechanical energy acts on the protein molecules of the actin and myosin filaments. Energy-rich phosphate compounds, which change from a high-energy state to a low-energy state during the chemical reactions, are the immediate sources of energy for muscle contraction. Some of these sources of energy include acetylcholine (ACh), ATP, and adenosine diphosphate (ADP). However, ATP, which acts as a reservoir of readily available energy, is the primary source of energy in muscle contraction. The regeneration of high-energy phosphates in muscles takes place continuously so that the supply of energy available to the muscles is not diminished.

The ATP needed for muscle work can be produced from aerobic or anaerobic metabolism. The aerobic metabolism of nutrients occurs in the oxidation of glucose or glycogen molecules and fatty acids to create ATP; this process is referred to as aerobic glycolysis. This metabolic process requires a continuous supply of blood in order to receive the required oxygen and nutrients to drive the process. Likewise, the muscles can produce ATP, work, or energy without oxygen, although the level of work produced is substantially less than that produced under aerobic conditions. In anaerobic conditions, the most immediate source of energy is through the production of the ATP molecule by breaking high-energy phosphate bonds in the CP molecule. The CP molecule contributes a phosphate (P) to an ADP molecule to create an ATP molecule and the required energy.

4.4.5 TYPES OF MUSCULAR WORK

Muscular work in occupational activities can be broadly categorized into groups: dynamic muscle work or static work. Dynamic muscular work involves coordinated groups of muscles working simultaneously to perform a movement in task execution. This type of work allows the flow of nutrients and oxygen to the muscles performing the task and allows longer duration in task performance, as compared to static muscular work. Most tasks involve some form of dynamic muscular work, even if it is only walking or occasional movement of the trunk or upper limbs. Static loads exist anytime a muscle has to perform a task without the benefit of the rhythmic movements that refresh the muscular system. This type of task activity is found in many occupations, but it is most frequently seen as a muscular activity isolated to a component of the body, such as the arms in holding a load during manual material handling. Additionally, static tasks are found in most material handling, environments, the electronics industry, and in repair and maintenance tasks (Figure 4.4).

Although dynamic work is preferable to static, even dynamic muscular contractions over extended periods of time can result in fatigue and a reduction in task performance. The amount of time an individual can sustain dynamic work is dependent upon the intensity of the work, the level of fitness of the individual, and environmental factors. A greater intensity of work can be sustained only over a short duration. Thus, if a very intense activity is required, this element of the task

FIGURE 4.4 (From Flickr, Pushing and pulling, 2006, http://farm1.static.flickr. com/74/193681147_7ac774b2a5.jpg?v=0 [accessed September 30, 2009].)

should be limited in duration, and a rest or refresh period should be designed into the process. Astrand and Rodhal (1970) found that for an individual to maintain an 8 h work shift, the task should not average more than 33% of the worker's maximum capacity for that task.

While dynamic work will lead to fatigue over an extended period of time, static work over a relatively short duration will also result in fatigue. In addition to the short-term strain that the body experiences from static loading, long-term injuries can result if static work is regularly performed. Injuries that can occur from extended or intense static work generally impact the soft tissues and spine. The conditions that can result include

- Arthritis due to mechanical stress
- Inflammation of the tendons and tendon sheaths
- Symptoms of arthrosis, a chronic degeneration of the joints
- Muscle spasms and soreness
- Intervertebral disc troubles

Static work increases the pressure inside the muscle, which together with the mechanical compression occludes blood circulation partially or totally and, as a result, muscles become fatigued more easily than in dynamic work. A muscle that is performing strong static effort is not receiving fresh blood, sugar, or oxygen and, therefore, must depend upon its reserves. A heavy static effort compresses blood vessels, caused by the internal pressure of the muscle tissue so that blood no longer flows through the muscle. Furthermore, waste products such as lactic acid are not removed during a strong static effort, and thus accumulate, producing pain, muscular fatigue, and latent muscle soreness.

As previously stated, there is a static component in most lifting, material handling, and physically intensive tasks. Making short, intermittent exertions and shifting the load between several muscle groups are strategies used to reduce rapid fatigue in statically loaded postures. Examples of task actions that involve static work include (Kroemer et al., 1994; Helander, 1995)

- Bending the back either forward or sideways
- Holding or carrying loads in the arms
- Manipulations that require the arms to be held out horizontally or at levels above the shoulders
- Resting the body weight on one leg while the other works a pedal
- Standing in one place for long periods of time
- Pushing and pulling heavy objects
- Tilting the head strongly forward or backward for extended periods
- Raising the arms above the shoulders for long periods of time
- Exertion of force to balance a load

In the initial stages of job design, it is important to find ways to reduce the static component of a task, as this can reduce the likelihood of premature muscle fatigue, and ensure that productivity is not negatively impacted.

Excessive efforts repeated over long periods of time, whether static or dynamic, can result in moderate to intense muscular aches and pains. These aches and pains may involve not only the muscles, but also the soft tissues, which can ultimately lead to cumulative injuries in the ligaments, joints, and tendons. These injuries are referred to as musculoskeletal disorders and are cumulative in development. They will be discussed in Chapter 7.

Symptoms of muscle overstress can be categorized into two groups:

- Reversible musculoskeletal problems
  - Pains of weariness, mostly localized to strained muscles and tendons
  - Short-lived
  - Pains disappear as soon as the load is relieved
- Persistent musculoskeletal problems
  - Localized to strained muscles and tendons
  - Affects the joints and adjacent soft tissues
  - Pains do not disappear when workload is relieved

Musculoskeletal pain and discomfort that is continually present over a number of years can lead to chronic inflammations of tendons and tendon sheaths, and deformation of joints. These types of conditions can lead to permanent disabilities and debilitating pain.

Epidemiological data indicates that maintaining static repetitive loads lead to a higher risk of the following:

- Inflammation of the tendons and possibly tendon sheaths (tendinitis or tenosynovitis)
- Inflammation of the attachment points of tendons
- Symptoms of arthrosis (chronic degeneration of the joints)

- Painful muscle spasms
- Sprains and strains
- Arthritis of the joints due to mechanical stress
- Intervertebral disc troubles

Although most of these conditions may be temporary, the continual exposure to risk factors can lead to permanent damage. The key is to recognize that occasional static muscular exertion in task performance is acceptable; however, the objective in task design should be to minimize the frequency and duration of these types of muscular contractions.

4.4.6 Types of Muscle Fibers in Muscular Contraction

Different muscles contain different types of muscle fibers. Some fibers contract rapidly, are glycolytic, and fatigue rapidly; these are known as fast twitch fibers. Other fibers are slowly contracting, oxidative, and slowly fatiguing and these are known as slow twitch fibers. Then there are fibers that are essentially a combination of these two types of fibers as they can contract rapidly and are both oxidative and glycolytic. Although a motor unit consists of only one kind of muscle fiber, most muscles are mixtures of fast and slow twitch fibers.

The slow twitch fibers are red in color due to an abundance of capillaries to provide oxygenated blood for the long-duration activities these muscles perform. Fast twitch muscle fibers have considerably fewer capillaries and thus tend to be white in color.

These two types of muscle fibers receive different types of innervation. Fibers in red muscles are innervated by motor neurons of small diameter, thus have a lower conduction velocity that discharge nearly continuously at low frequency. Fibers in white muscles receive innervation from larger motor neurons that have longer periods of silence between discharges, but discharge at high frequencies.

The properties of both slow and fast twitch muscle fibers are summarized in Table 4.1. The properties of slow muscle fibers make them most suited for extended periods of contraction where a minimum force is required, for example, in the maintenance of posture. Fast twitch muscle fibers are better suited to short periods of rapid contraction at higher forces, for example, in sprint running. Strength training leads to hypertrophy of mainly white muscles and while the number of fibers does not increase, the size of the fibers and the number of myofibrils do increase. This increases both the strength and velocity of contraction.

4.4.7 Muscular Fatigue

During prolonged, repetitive, or intense contraction, muscle fatigue can occur. Studies on athletes during prolonged submaximal exercise have shown that muscle fatigue increases in almost direct proportion to the rate of muscle glycogen depletion. This can happen in static or dynamic muscular movement; however, the onset of fatigue is much faster in static contractions. The rapid fatigue during static contraction is due to the lack of nutrients refreshing the muscles and an accumulation of waste byproducts, such as

TABLE 4.1

Properties of Fast and Slow Twitch Muscle Fibers

Property	Fast Twitch Muscle Fibers (Type II)	Slow twitch Muscle Fibers (Type I)
Example		Soleus (calf) muscle
Twitch contraction time	Fast	Slow
General color	White	Red
Capillary blood flow	Low	High
Myoglobin content	Low	High
Primary source of ATP	Glycolysis	Oxidative phosphorylation
Glycogen	High	Low
Tension/force produced	Larger	Smaller
Myosin-ATP activity	High	Low
Resistance to fatigue	Low	High
Nerve fiber size	Large	Small
Nerve fiber activity	Intermittent, high frequency	Continuous, low frequency
Oxidative capacity	Low	High

lactic acid. Also, during vigorous dynamic exercise, the circulatory system cannot supply oxygen to muscle fibers quickly enough. In the absence of oxygen, the muscle cells begin to produce lactic acid, which accumulates in the muscle. The lactic acid buildup lowers pH, and as a result, muscle fibers no longer respond to stimulation.

To reduce the likelihood and occurrence of muscular fatigue in occupations, the task should be designed such that submaximal strength requirements are expected, and dynamic muscular activities are the primary task activity. Additionally, work-rest cycles should be integrated into the process by adding breaks to the task, job sharing (two or more operators performing a task), or task variation (allow the operator to do multiple tasks during a complete cycle of task performance).

It is important to effectively address muscular fatigue and recognize the impact of fatigue on task performance, especially when it is premature or occurs frequently. Studies have shown that muscular fatigue can be an indicator of poor task design and lead to occupational injuries; thus, addressing this seemingly benign risk factor can be effective in promoting a more efficient and safe work environment.

4.4.8 Types of Muscle Contractions

The types of muscle contractions include five categories of contractile functionality. The types of muscle contractions include isometric, isoinertial, eccentric, isotonic, and isokinetic.

4.4.8.1 Isometric Contractions (Static Muscular Contraction)

An isometric contraction is one in which the muscle is activated, but instead of being allowed to lengthen or shorten, it is held at a constant length. An example of an isometric contraction would be carrying an object in front of the body. The weight of the object would be pulling downward, but the hands and arms would be opposing the motion with equal force going upward. Since the arms are neither raising nor lowering, the biceps will be performing an isometric contraction.

4.4.8.2 Isoinertial Contractions (Constant Load)

In isoinertial contractions a constant load is applied to the muscles. In this type of muscular contraction, the muscles are responding to a constant load where the measurement system considers acceleration and velocity.

4.4.8.3 Eccentric Contractions (Lengthening of the Muscle)

Eccentric contraction causes a muscle to lengthen under tension. Such contractions are used to resist external forces such as gravity. The quadriceps muscles, for instance, undergo eccentric contractions when a person walks down steps, runs downhill, or lowers a weight. Eccentric contractions also occur during the deceleration phases of running. The downward phase of a biceps curl, for example, requires eccentric action of the biceps muscle.

4.4.8.4 Isotonic Contractions (Concentric Shortening)

In isotonic contractions, the muscle contracts and shortens, giving movement. Isotonic contractions involve the muscle in a situation where an equal amount of tension is being developed in the muscle throughout the exercise. The muscle develops equal tension while the muscle length changes. Some examples are pull-ups, push-ups, and lifting weights.

4.4.8.5 Isokinetic Contractions (Constant Force)

In isokinetic contractions, the muscles exert a constant force. An isokinetic muscle contraction is one in which the muscle contracts and shortens at a constant rate of speed. This type of muscle contraction usually requires special, training equipment that increases the load as it senses that the muscle contraction is speeding up.

4.4.9 MEASUREMENT OF MUSCULAR STRENGTH

The measurement of muscular strength is a reliable approach to assess human capability and limitations related to task performance, product design, or recreational activities. A number of factors should be considered when designing an approach to measure muscular strength, including the objective of strength testing (i.e., task evaluation, product design, or risk assessment), muscular requirements, costs, and types of task activities. In ergonomics, the most common strength-testing methodologies are three broad categories that include isometric, dynamic (isoinertial and isokinetic), and psychophysical testing (Gallagher et al., 1998; Kroemer, 2006b.)

4.4.9.1 Isometric Strength Testing

Isometric strength is defined as the capacity to produce force or torque with a voluntary muscular contraction while maintaining constant muscular length (Gallagher,

1998). Isometric strength-testing equipment has not been standardized, and a variety of products are available to assess static strength, given the muscles being evaluated. This is perhaps the most studied and measured strength-testing approach, as it is comparatively easy to conduct, measure, and understand. The ease of use and reduced likelihood for injury are advantages associated with isometric testing. However, a primary criticism of this method is that these static strength tests are not compatible with most task performance, given the dynamic nature of most jobs particularly those involving material handling. Nonetheless, this is a widely accepted approach for strength assessment in applied and research ergonomics, due to the historic success in applying results of static tests to task evaluation, injury prediction, and design.

4.4.9.2   Dynamic Strength Testing

Dynamic strength tests include any type of assessment that allows the movement of the muscle during the testing process. Isoinertial strength assessment is done when mass properties of an object remain constant during the dynamic muscular exertion, such as a task where a weight is lifted over a predetermined distance. Isokinetic strength testing maintains a constant load throughout a predetermined range of motion.

4.4.9.3   Psychophysical Testing

The psychophysical assessment for determining maximum dynamic lifting capacities has been used in physiological research for many years. The term "psychophysical" refers to a methodology in which subjects perform lifts to determine maximum, safe levels of exertion based on their personal perspective of acceptable risk. In practice, this method is dependent upon cooperation from the subject, expertise from the evaluator, and good judgment on the part of both the subject and evaluator to ensure the safety of the test and the accuracy of the physical performance data.

4.4.9.4   Procedure for Strength Testing

Regardless of whether or not the test is dynamic or static, the basic factors to consider when designing a process for strength testing are as follows (Gallagher et al., 1998):

- Equipment available to make the measurements
- Instructions to give to subjects being tested
- Duration of measurement period
- Duration of rest period
- Body posture (torso, arms, legs) during test
- Number of trials
- Physical condition of subject
- Type of postural control during test
- Environmental conditions during test
- Subject motivation

The initial protocol for isometric strength testing was developed by Caldwell et al. (1974) and adapted in succeeding years as research revealed needed adjustments to the approach. Although this test approach was designed for static testing, it can be adapted for dynamic testing as well. The procedure from the Caldwell protocol is described as follows (Plog and Quinlan, 2002) for the measurement of static strength:

- Determine type of static muscular exertion to be measured.
  - Ensure relevance of measured to intended research or application.
- Measure static strength as follows:
  - Assess static strength during a steady exertion for 4 s.
  - Disregard transient periods of 1 s before and 1 s after exertion.
  - The strength datum is the mean score recorded during the first 3 s of the steady exertion.
- Treat subjects as follows:
  - Inform subject about the purpose of the test, equipment, and procedures.
  - Provide detailed and factual instructions without emotion or coaching.
  - Instruct subject to "increase maximal exertion (without jerk) in one second and then maintain this effort during a four second period."
  - A different time and procedure may be utilized for special conditions, such as obtaining finger strength.
  - Provide feedback to subject during the test in a qualitative, non-comparative, positive manner. Do not give instantaneous feedback during the exertion period.
  - Avoid incentives that can affect subject's motivation such as rewards, goal setting, competition, spectators, fear, and environmental factors (i.e., noise, heat).
- Provide a minimal rest period of 2 min between related efforts and exertions, more if symptoms of fatigue are apparent.
- Describe the conditions that will exist during strength testing as follows:
  - Body parts and muscles primarily used.
  - Body posture and movement.
  - Body support or reaction force available.
  - Coupling of the subject to measuring device.
  - Strength measurement and recording device.
- Describe the following subjects:
  - Population and sample selection including sample size.
  - General health (survey or medical examination recommended).
  - Gender.
  - Relevant anthropometrics.
  - Training related to strength testing.
  - Occupation.
- Report results of experiment.

The analysis of data on strength testing should include calculation of summary statistics, distribution of data, and statistical analysis to evaluate relevant hypotheses that can be useful in assessing the compatibility of the strength values with the intended population. The data to be reported includes the following at a minimum:

- Measures of central tendency (mean, media, mode)
- Measures of variation (standard deviation, variance, and range)

- Identification of outliers
- Skewness of data
- Graphical representation of data (i.e., histogram)
- Percentiles

Upon completion of data collection and analysis, the results are then ready to be used for making inferences about the subject population.

4.5 SUMMARY

Muscles skeletal and nervous systems are the integrated systems that permit movement of the human body. They convert chemical energy, obtained from food and drinks, into mechanical energy, which creates force and allows movement of the body limbs and activities that result in task performance. Designing occupational environments, equipment, and processes to be compatible with the musculoskeletal system is a goal of ergonomics. Understanding muscle work and how stimuli are processed by the body is imperative for the ergonomist. How multiple bodily functions interact to allow task performance requires knowledge of not only the musculoskeletal system but the nervous system as well. The nervous system controls the musculoskeletal system via motor and sensory nerves, which, in turn, communicate with the central nervous system. Additionally, understanding the science behind the transformation of energy into muscle movement and also the factors contributing to fatigue will aid in efficient and safe task design.

Case Study

Ergonomic Analysis of a Telemarketing Operation
By Robert O. Andres, David D. Wood, and Nancy E. Laurie

Introduction

This report presents the results of our ergonomic analysis a telemarketing operation in the call center of a large mail order firm. Dr. Andres first observed the telemarketing floor during his visit on November 5, 1993. A subsequent site visit took place May 4, 1994; ergonomists Nancy Laurie, M.S. and David Wood, M.S.I.E. accompanied Dr. Andres and performed the analyses summarized in this report. It has been brought to our attention that since our last visit a possible change is being considered—the supervisors would be distributed about the floor so that the lead station and supervisors' offices would no longer be centralized. The implications of this change will be discussed in the Conclusions and Recommendations section.

Methods

The site visit on 5/4/94 yielded several types of data. Video records of task performance were acquired, both to document worker technique and worker interaction with their workstation as well as to provide examples that can be used during training (if training is pursued as an intervention). We measured the dimensions of the existing workstations across the entire floor area to determine what is available. We were also supplied with OSHA 200 logs so that we could understand the nature of the injuries and illnesses experienced by workers overall and in the telemarketing area.

Clerical worker exposure to ergonomic risk factors was estimated from both the on-site visit and a video tape analysis in our lab. Observational data and video data were summarized using ANSI's Z-365 checklist. The OSHA 200 Logs were reviewed; only ergonomic injuries were summarized (cumulative trauma disorders [CTDs]).

Results

There were three lost time ergonomic injuries in 1993 in the clerical operation. We estimated the severity rate using the following formulas (Bureau of Labor Statistics [BLS]):

Total hours per year = 415 (workers) × 30 (h/week) × 50 (weeks/year) 622,500 h/year

Severity rate for lost days = number of lostdays/200,000 h/year (57[lost days]

× 200,000)/622,500 = 18.31 lost days per 100 person years

Incidence rate for lost time injuries = number of lost time injuries/200,000 h/year (3

× 200,000) / 622,500 = 0.96 incidents per 100 person years

There were nine other lifting injuries; each had no associated lost time. There were 268 lost days due to three slip/fall incidents. Seventeen other falls had no associated lost time. We gave a quick glance at the warehouse OSHA 200 Log ergonomic injuries. In absolute terms the number of lost days was some 10–12 times higher than those in telemarketing. However, without knowing the number of hours spent working last year in the warehouse, the incidence and severity rates could not be calculated.

After the site visit and viewing the video shot during our visit an ANSI Z-365 general ergonomic risk assessment checklist was filled out. Two aspects of the telemarketing operator's work (mechanical pressure at wrist and neck flexion) were above the trigger point (meaning that engineering or administrative controls should be implemented to reduce the physical stress on the workers). The workers rest their wrists on whatever surface is available as they type. This mechanical pressure is a risk factor for CTDs of the hand/wrist. The neck is also a potential site for CTDs. Workers must continuously look down at forms and papers on their desks and then look up at their computer monitor.

Table 4.2 contains a summary of the workstation heights. Only workstations with a computer and chair were included; the actual number of workstations used varies over time with the volume of sales.

TABLE 4.2

Workstation Height Summary

Workstation Height (in.)	Number
26.50	11
26.75	154
27.00	71
28.00	90
28.25	18
30.25	3
30.50	59
30.75	1
31.00	4
Total	411

Conclusions and Recommendations

The number of CTDs found on the OSHA 200 logs was small (3). Comparing the incidence and number of lost workdays with nationally published values for the finance, real estate, and insurance sectors (which are the most clerical of the classifications published by the Bureau of Labor Statistics in Accident Facts, 1993 Edition, referring to the 1991 statistics), the national average incidence of lost workday rate was 1.1 (versus 0.96 at here) while the average lost workdays were 24.1 (versus 18.3 at here). The clerical operation is better than the national average for the number of lost workday cases and for the number of lost workdays. That does not mean that improvements cannot be made.

Suggested improvements will be proposed on both a short term and a long term basis. The short term suggestions are generally administrative controls, whereas the long term suggestions involve engineering controls as well.

Short term improvements

Assign workers to proper work surfaces based on height:

We understand that you are considering dividing the floor into teams; this may make assignment more complicated but we will provide worker heights matched to work surface heights if you elect to pursue this approach.

Make sure that footrests are placed where needed:

We noticed that footrests were not present in all workstations, and we also noticed that they were pushed to the side and not used in several situations where the worker was too tall for their work surface.

Provide palm rests for the workers with wrist/hand complaints:

This may involve getting more palm rests because you do not want to take them away from people that already have them (we understand that this did happen at some locations on the floor as a result of the December 1, 1993 incident, and many workers expressed their dismay that they had been taken away).

We do not believe that all workers need these, but for those that experience some discomfort or have developed some early symptoms (see #6 below) they could be provided on a case by case basis.

Train the workers to properly adjust their chairs and to sit properly:

The chairs that the majority of workers sit in seem to be adequate ergonomically. However, a good chair set in the wrong position can still cause problems during prolonged sitting. Training should be provided on how to adjust the chairs, and why. For example, if discomfort develops in the low back, the seat back may be too high, or if the back of the legs get sore then the seat pan may be set too high. Proper sitting posture should also be taught.

Train the workers how to stretch properly, and provide them with stretch breaks:

We have worked with clients who have implemented stretch break programs ranging from voluntary when the worker feels the need to mandatory 5 min every hour. The stretches need to be appropriate and they need to be either led by properly trained leaders or the individual workers need to receive training. We have developed a stretch guide for workers that was designed to fulfill this need. Implement a medical management program that would include training for the workers and supervisors so that they can recognize early symptoms of CTD problems, and make sure that management promotes a policy of early reporting so that CTDs are prevented:

This type of program would need to be corporate-wide to be effective. Our experience indicates that an effective medical management program can prevent minor aches and pains from becoming lost workday incidents. You can make all the biomechanical modifications possible and still have a large number of lost days if early reporting and aggressive pursuit of conservative treatment are not part of corporate policy.

Long term improvements

Provide better work surfaces with rounded front edges:

The sharp front edges on the existing stations can cause mechanical stress concentrations on the tendons of the forearms if workers rest their arms on them.

Investigate adjustable work surfaces:

These adjustable surfaces are rapidly coming down in price. In future facilities, it may be cost effective to buy a percentage of adjustable work surfaces to adapt to anthropometric extremes. This will become more cost effective if the company moves to more multiple shifts sharing the same stations.

Provide a small number of standing work stations:

These standing stations could be placed along a side wall, and as long as the workers can maintain their identity by logging on to one of these stations, they could be used for up to 20 min at a time on a voluntary basis for workers that tire from continuous sitting.

Provide monitor stands:

Although the monitor heights in the current workstations appear adequate, as workers are trained to better adjust their chair to their worksurface height to promote the best body postures for keyboard work and writing, the need for adjustment of monitor height will increase. These devices will be needed most by the taller workers. These devices will also help if multiple shifts will be using the same workstation. However, the nature of the work performed should be taken into account—you do not want to raise all of the monitors if it will force people to look up and down more.

Before concluding this report, the lifting and slip and fall incidents need to be addressed. Even though lifting is not a primary component of the jobs performed by those answering the phones (and perhaps because it occurs so infrequently), there were nine incidents due to lifting in clerical employees. These employees need to be trained in proper lifting mechanics. The prevention of slips and falls for these workers is more a safety issue than an ergonomic issue, but slips and falls often lead to back injuries. Three lost day incidents were caused by falls experienced by clerical workers. Seventeen other falls or trips were experienced by office workers, but these did not lead to lost days. Although the injury log did not indicate where the falls took place, most of them may have taken place in another location besides the Telemarketing floor. Disparate surfaces (like tile and carpet or concrete and ice) can cause slips or trips when people move from one surface to the other. Obvious trip hazards in the telemarketing area should be removed if found.

The implementation of any of these short or long term interventions could best be handled by an active Ergonomics Committee as part of a structured Ergonomics Program. Such a program would include the training for the managers, supervisors, and employees that is referred to above, as well as generally raising the level of awareness of ergonomic issues across the entire telemarketing operation. With the approaching Ergonomics Standard being developed by OSHA, it would be better to start this process sooner instead of later.

In summary, even though the telemarketing operation currently has incidence and severity rates below the national average, several suggestions for improvements have been presented. Based on the OSHA logs we reviewed, however, it appears that the maximum opportunity for cost justifying ergonomic interventions is found in the Fulfillment Center distribution operation.

Source: Total contents of Case Study reprinted with permission from Ergoweb.com

---

EXERCISES

4.1 Define muscular work.

4.2 Explain how energy (work) can be exerted without actually performing work as defined by laws of physics.

4.3 Explain the muscle contraction process.

4.4 Explain the chemical processes that take place in aerobic an anaerobic contraction.

4.5 Define braches of the peripheral nervous system that innervate the muscles.

4.6 What is the difference in a reflex and a muscular movement such as kicking a ball?

4.7 Explain why muscles fatigue more quickly in static loading. Identify an occupational task where static loading is required.
4.8 Discuss the different types of muscle contractions.
4.9 Describe a process for dynamic strength testing.
4.10 Develop a process for collecting static strength.

REFERENCES

Asimov, I. (1989). *Asimov's Chronology of Science and Discovery*, Harper & Row: New York.
Astrand, P.O. and Rodahl, K. (1970). *Textbook of Work Physiology*, McGraw-Hill: New York.
Astrand, O. and Rodahl, K. (1977). *Textbook of Work Physiology, Physiological Bases of Exercise*, 2nd edn., McGraw-Hill: New York.
Caldwell, L.S., Chaffin, D.B., Dukes-Dobos, F.N., Kroemer, K.H., Laubach, L.L., Snook, S.H., Wasserman, D.E. (1974). A proposed standard procedure for static muscle strength testing. *The American Industrial Hygiene Association Journal*, 35(4), 201–206.
Culver Academies. (n.d.). The eight characteristics of life. http://academies.culver.org/science/kinseyk/nervoussystematlas.gif (accessed September 30, 2009).
Darling, D. (2009). The Internet encyclopedia of science. http://www.daviddarling.info/images/rotator_cuff.jpg (accessed September 30, 2009).
Flickr. (2006). Pushing and pulling. http://farm1.static.flickr.com/74/193681147_7ac774b2a5.jpg?v=0 (accessed September 30, 2009).
Gallagher, S. (1991). Acceptable weights and physiological costs of performing combined manual handling tasks in restricted postures. *Ergonomics*, 34(7), 939–952.
Gray, H. (1918). *Anatomy of the Human Body*, 20th edn., Lea & Febiger: Philadelphia, PA. Retrieved September 9, 2009, from http://www.bartleby.com/107/
Grey, G. (2008). The science of sport. http://www.cbc.ca/canada/british-columbia/features/athletesblog/2008/06/george_grey_the_science_of_tra.html (accessed September 30, 2009).
Helander, M. (1995). *Guide to Ergonomics of Manufacturing*, 2nd edn., CRC Press: New York.
Hockenbury, D. and Hockenbury, S. (2008). *Psychology*, MacMillan Publishers: New York.
Huxley, H. and Hanson, J. (1954). Changes in the cross-striations of muscle during contraction and stretch and their structural interpretation. *Nature*, 173(4412), 973–976. DOI:10.1038/173973a0. PMID 13165698.
Huxley, A.F. and Niedergerke, R. (1954). Structural changes in muscle during contraction: Interference microscopy of living muscle fibres. *Nature*, 173(4412), 971–973. DOI:10.1038/173971a0. PMID 13165697.
Kroemer, K. (2006a). Engineering anthropometry, *The Occupational Ergonomics Handbook*, 2nd edn. (Eds. W.S. Marras and W. Karwowski), Chapter 9, CRC Press–Taylor & Francis: Boca Raton, FL, pp. 9.1–9.30.
Kroemer, K. (2006b). Human strength evaluation, *The Occupational Ergonomics Handbook*, 2nd edn. (Eds. W.S. Marras and W. Karwowski), Chapter 10, CRC Press–Taylor & Francis: Boca Raton, FL, pp. 10.1–10.23.
Kroemer, K.H.E. and Grandjean, E. (1997). *Fitting the Task to the Human: A Textbook of Occupational Ergonomics*, 5th edn., Taylor & Francis: New York.
Kroemer, K.H.E., Kroemer, H.B., and Kroemer-Elbert, K.E. (1994). *Ergonomics: How to Design for Ease and Efficiency*, Prentice Hall: Englewood Cliffs, NJ.
Magnus, D. and Cho, M.K. (2005, June 17). Issues in oocyte donation for stem cell research. *Science*, 308(5729), 1747–1748. DOI: 10.1126/science.1114454. Published Online May 19, 2005. http://www.sciencemag.org/content/308/5729/1747.full
Medical-Look. (2009). Skeletal muscle fiber. http://www.medical-look.com/systems_images/Skeletal_Muscle_Fibers.gif (accessed September 30, 2009).

Pancaldi, G. (2003). *Volta*, Princeton University Press: Princeton, NJ.

Plog, B.A. and Quinlan, P.J. (2002). *Fundamentals of Industrial Hygiene*, 5th edn. (Eds. B. Plog and P. Quinlan), United States National Safety Council: Itasca, IL, pp. 3–32.

Prada-Leon, L.R., Celis, A., and Avila-Chaurand, R. (2005). Occupational lifting tasks as a risk factor in low back pain: A case-control study in a Mexican population. *Work*, 25(2), 107–114.

Program BPPT. (n.d.). Biologi. http://www.pspnperak.edu.my/biologit5/Abd%20Razak%20 b.%20Yaacob/Portfolio/BBM/Audio/saraf/sensory_neuron_jpg.jpg (accessed September 30, 2009).

Reader's Digest. (2009). How to make walking a habit. http://media.rd.com/rd/images/rdc/ mag0904/how-to-make-walking-a-habit-af.jpg (accessed September 30, 2009).

Rogers, G. (2004). Windsurfing fitness. http://www.boardseekermag.com/windsurfing_fitness/ pics/static_hold.jpg (accessed September 30, 2009).

Spiliotis Lab. (2009). Drexel University. http://www.pages.drexel.edu/~ets33/research.html (accessed September 30, 2009).

Takken, T., Ribbink, A., Heneweer, H., Moolenaar, H., and Wittink, H. (2009). Workload demand in police officers during mountain bike patrols. *Ergonomics*, 52(2), 245–250.

Tangient LLC. (2009). FRF muscular system. https://eapbiofield.wikispaces.com/ FRF+Muscular+System?f=print (accessed September 30, 2009).

TopNews. (2009). Spinal cord. http://topnews.us/images/spinal-cord.jpg (accessed September 30, 2009).

United States Department of Agriculture (USDA). (2007). Agricultural research service: Ergonomic disorders. http://www.ars.usda.gov/sp2userfiles/ad_hoc/19000000SafetyH ealthandEnvironmentalTraining/graphics/EgDequervains.jpg (accessed September 30, 2009).

University of Virginia: School of Medicine. (2009). Med-ed. http://www.med-ed.virginia.edu/ courses/cell/handouts/images/Muscle1.jpg (accessed September 30, 2009).

U.S. National Library of Medicine. (2009). Genetics home reference—Actin. http://ghr.nlm. nih.gov/handbook/illustrations/actin.jpg (accessed September 30, 2009).

Verma, S.K., Lombardi, D., Chang, W.-R., Courtney, T.K., and Brennan, M.J. (2008). A matched case-control study of circumstances of occupational same-level falls and risk of wrist, ankle and hip fracture in women over 45 years of age. *Ergonomics*, 51(12), 1960–1972. http://dx.doi.org/10.1080/00140130802558987 (accessed July 6, 2009).

Wikimedia Commons. (2009). Robbie McEwen 2007 Bay Cycling Classic. http://upload.wiki- media.org/wikipedia/commons/f/f4/Robbie_McEwen_2007_Bay_Cycling_Classic_2. jpg (accessed September 30, 2009).

Williams, H.S. (2009). Modern development of the chemical and biological sciences. Retrieved January 21, 2011, from http://embryology.med.unsw.edu.au/history/page4.htm

5 Anthropometry

5.1 LEARNING GOALS

The objective of this chapter is to introduce the study of anthropometrics and teach principles that can be used to support anthropometric design. Anthropometric history, terminology and application is presented. The process for data collection, calculation of statistics, and identification of appropriate data sources will also be covered.

5.2 KEY TOPICS

- History of anthropometry
- Anthropometric design principles
- Principles of universal design
- The myth of the "average" human
- Anthropometric terminology
- Anthropometric measurement
- Static and dynamic dimensions
- Variability
- Anthropometric data
- Anthropometric tools
- Anthropometric data sources
- Design using anthropometric data
- Anthropometrics and biomechanics
- Anthropometric application

5.3 INTRODUCTION AND BACKGROUND

Prior to World War II, most of the work that was intended to be ergonomic was primarily focused on tests for selecting the proper people for jobs and on the development of improved training procedures. The focus was clearly on fitting the person to the job, which is the opposite of the goal of ergonomics to: "Fit the task to the person." However, growth of the field in the ergonomics occurred near the end of World War II in 1945 when engineering psychological laboratories were established by the U.S. Air Force and U.S. Navy. This was the result of observing that no matter how much training and job selection went into task performance, the complex equipment still exceeded the capabilities of the people who had to operate it. Thus, they were forced to reconsider fitting the equipment to the person. This was also known as the "knob and dial" era. Much research was conducted during this time to develop optimal design parameters for displays and controls.

The word *anthropometry*, from anthro (man) and meter (measure), is defined as the "measurement and study of the human body and its parts and capacities"

(WorldNet, 2011). An early use of anthropometry is recorded by the savant Alphonse Bertillon in 1883 for criminal record purposes. He found that by collecting a series of specific measurements like the length of a middle finger, or the width of the head from temple to temple, he could consistently distinguish individuals from one another without error (Sutherland, 1937). By 1894, this method was replaced by finger printing, at the recommendation of Francis Galton, who was, incidentally, a cousin of Charles Darwin (O'Connor, 2011). During the 1930s, anthropometric approaches were used in Nazi Germany in order to determine which races were "inferior" and, more importantly, to distinguish Aryans from Jews (MSNBC).

Today anthropometry is used in a variety of fields including engineering, anthropology, ergonomics, and physical therapy. For instance, anthropometrists can study correlations in the body dimensions of different races based on human evolution and migratory patterns of ancient hominids. Human characteristics, like physical stature, can demonstrate the health of a society. According to such studies, American soldiers during the Revolutionary War were, on average, 2 in. taller than their British counterparts; this suggests that the New World was a better socioeconomic environment for humans at the time. Another historical example of generational or societal body dimension differences is that the average weight of a student entering West Point Military Academy in antebellum America was under 128 lb, and 25% were under 110 lb—a reflection on the general socioeconomic well-being of the general population. Studies also show that illiterate men were smaller in stature than literate men (Komlos, 1992).

Another common, modern use of anthropometry is for product optimization. Clothing companies consult anthropometric trends to determine the distributions of sizes clothing to manufacture. Just recently, for instance, in response to what seems to be a first world obesity epidemic, a British retailer for plus-sized clothing has released the largest clothing size yet: XXXXL, four Xs. This type of anthropometric phenomenon would not likely have occurred at any other point in human history (Huff, 2010).

The three primary factors that differentiate humans in measurement are gender, ethnicity, and age (Wang etal., 2006). Large differences in body size due to gender and genetics are common. For instance, men are, on average, 13 cm (5 in.) taller than women, as well as larger in most other body measures, particularly in upper body strength capabilities. Ethnicity differences have been found when comparing individuals living in different countries. For example, the average male stature in the United States is 167 cm (66 in.), while the average male stature in Vietnam is 152 cm (60 in.). For these reasons, it is important to understand the process of anthropometric data collection, analysis, and application.

5.4 WHAT IS ANTHROPOMETRY?

The study of anthropometrics, or human measurement, is concerned with the physical sizes and shapes of humans. Anthropometrics is a very important branch of ergonomics (Pheasant, 1996) in research and application. This science deals with the measurement of size, mass, shape, and inertial properties of the human body. Anthropometry relies on sophisticated methods to measure physical dimensions including static and dynamic measurements of specific populations. The results obtained from these methods are statistical data that can be applied in the design of products, clothing, occupational, and recreational environments. Also, anthropometric data is essential in developing biomechanical models to predict human movement, reach, force, and space requirements.

Static or structural anthropometry is focused on specific skeletal dimensions when the body is not in motion, while dynamic or functional anthropometry measures distances and ranges of motion while the body is moving or involved in physical activities. Anthropometric design principles should be applied in design for a large cross section of the population to promote physical comfort in the workplace, recreational, and service environments. Ignoring these physical requirements can create bad postures that lead to fatigue, loss of productivity, and sometimes injury. Furthermore, anthropometry not only deals with establishing appropriate working heights, but also facilitating access to controls and input devices for users.

Anthropometric data has many uses and can be applied in a variety of industries. For instance, furniture designers use anthropometric data to accommodate the variations of a wide range of end users. Manufacturing facilities can use anthropometrics to design workstations for their employees. The military uses anthropometric data to design equipment for soldiers. Aircraft manufacturers use anthropometrics to design passenger seating. Anthropometric data can be used in the medical device field to develop prostheses for a particular ethnicity. Anthropologists can also use anthropometric data to compare the effect of nutrition on different cultures based on body dimensions. Typical anthropometric measurements and associated percentiles are contained in the appendix.

5.4.1 ANTHROPOMETRIC DESIGN PRINCIPLES

The following three principles provide guidelines for using anthropometric data in design. The principles can all be used in a single design and the degree to which each is applied will depend on the user population, consequence for lack of accommodation, and economic factors of the design. For examples of population anthropometric percentiles see Figures 5.1 and 5.2.

5.4.1.1 General Guidelines

1. Design for adjustable ranges
   a. Provide a range of adjustments, particularly when health and safety issues are involved (i.e., driving a car, working in office).
   b. Mitigate the likelihood of eliminating sections of the user population due to design limitations, by increasing ranges through adjustability.
2. Design for extremes
   a. Design for *maximum* value of design features (i.e., height of doors, size of buttons) or design for *minimum* value of design features (i.e., distance from controls).
   b. Traditionally, extremes have been addressed by designing for
      i. 5th percentile female
      ii. 95th percentile male

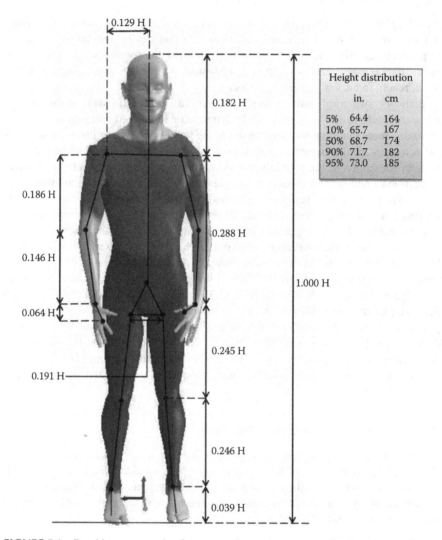

FIGURE 5.1   Provides an example of stature and seated mesurment for American male.

3. Design for average users
   a. Use 50th percentile figures for relevant dimensions.
   b. This strategy is only acceptable when you are primarily concerned with one dimension and health and safety is not a significant issue (i.e., auditorium seats); legislations such as the Americans with Disabilities Act (ADA) require additional accommodation in some environments.

5.4.1.2 Principles of Universal Design

Universal design is defined as designing products, buildings, and exterior spaces to be usable by all people to the greatest extent possible. The use of anthropometric

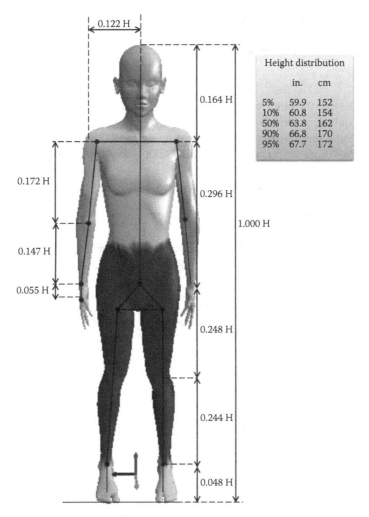

FIGURE 5.2 Provides an example of stature and seated measurement for a 40 year old Japanese American female.

data to accommodate a global society benefits from universal design principles. These concepts were developed by a working group of architects, product designers, engineers, and environmental design researchers. This working group collaborated to establish broad principles that can be used in the design of environments, products, and communications (Mace et al., 1996).

These seven principles may also be applied to the assessment of existing designs, development of instructional material, and in consumers' evaluations of products; however, it should be pointed out that all guidelines may not be relevant to all designs. The Center for Universal Design at North Carolina State University lists the design principles and associated guidelines as follows (CUD, NCSU, 2011):

- **Principle 1: Equitable Use**

The design is useful and marketable to people with diverse abilities.

Guidelines

    1a.   Provide the same means of use for all users, identical whenever possible, equivalent when not.
    1b.   Avoid segregating or stigmatizing any users.
    1c.   Provisions for privacy, security, and safety should be equally available to all users.
    1d.   Make the design appealing to all users.

- **Principle 2: Flexibility in Use**

The design accommodates a wide range of individual preferences and abilities.

Guidelines

    2a.   Provide choice in methods of use.
    2b.   Accommodate right- or left-handed access and use.
    2c.   Facilitate the user's accuracy and precision.
    2d.   Provide adaptability to the user's pace.

- **Principle 3: Simple and Intuitive Use**

Use of the design is easy to understand, regardless of the user's experience, knowledge, language skills, or current concentration level.

Guidelines

    3a.   Eliminate unnecessary complexity.
    3b.   Be consistent with user expectations and intuition.
    3c.   Accommodate a wide range of literacy and language skills.
    3d.   Arrange information in order of its importance.
    3e.   Provide effective prompting and feedback during and after task completion.

- **Principle 4: Perceptible Information**

The design communicates necessary information effectively to the user, regardless of ambient conditions or the user's sensory abilities.

Guidelines

    4a.   Use different sensory modes (visual (pictorial), auditory (verbal), tactile) for redundant presentation of essential information.

4b. Provide adequate contrast between essential information and supplemental information and backgrounds.
4c. Maximize "legibility" of essential information.
4d. Differentiate elements in ways that can be easily clarified (i.e., make it easy to give instructions or directions).
4e. Provide compatibility with a variety of techniques or devices used by people with sensory limitations.

- **Principle 5: Tolerance for Error**

The design minimizes hazards and the adverse consequences of accidental or unintended actions.

Guidelines

5a. Arrange elements to minimize hazards and errors; that is, make the most used elements the most accessible. Arrange the more hazardous elements in an isolated or shielded manner, or ideally, eliminate access to these elements.
5b. Provide warnings of hazards and errors.
5c. Provide fail-safe features.
5d. Discourage unconscious action in tasks that require vigilance.

- **Principle 6: Low Physical Effort**

The design can be used efficiently and comfortably, while minimizing fatigue.

Guidelines:

6a. Allow user to maintain a neutral body position.
6b. Use reasonable operating forces.
6c. Minimize repetitive actions.
6d. Minimize sustained physical effort.

- **Principle 7: Size and Space for Approach and Use**

Appropriate size and space is provided for approach, reach, manipulation, and use regardless of user's body size, posture, or mobility.

Guidelines

7a. Provide a clear line of sight to important elements for any seated or standing user.
7b. Design reach to all components accessible and comfortable for any seated or standing user.
7c. Accommodate variations in hand and grip size.
7d. Provide adequate space for the use of assistive devices or personal assistance.

5.4.1.3 Myth of the Average Human

The average user is often referred to as a mythical being, because there is no known person whose body dimensions per segment, all fall within the 50th percentile. Body dimensions are not linearly correlated so, for example, a person who has long arms does not necessarily mean that they will have a long neck or long legs. Thus, the goal to design for the average user should be abandoned and statistical ranges or percentiles should be used to guide design decisions.

5.4.2 ANTHROPOMETRIC TERMINOLOGY

In order to further understand traditional anthropometric measurement procedures and techniques, it is important to first define some specific terminology measuring conventions and definitions (Table 5.1).

5.4.2.1 Definitions of Measurements

The following terms are used in anthropometric discussion. The relevant definition is provided for each as it relates the measuring the human body.

- *Height* is a straight line, point-to-point vertical measurement, and *breadth* is a straight line, point-to-point horizontal measurement running across the body or a segment.
- *Depth* is a straight line, point-to-point horizontal measurement running fore and aft the body.
- *Distance* is a straight line, measurement between landmarks on the body.
- *Curvature* is a point-to-point measurement following a contour, and this measurement is typically neither closed nor circular.
- *Circumference* is a closed measurement that follows a body contour and, therefore, is not circular.
- *Reach* is a point-to-point measurement following the long axis of the arm or leg.

5.4.2.2 Anthropometric Planes

The three planes that are more commonly used in obtaining anthropometric measurements are the sagittal, coronal, and transverse planes (Figure 5.3). Each is described as follows:

- *Sagittal plane*—the plane dividing the body in the median plane. The mid-sagittal plane divides the body into equal left and right parts in the median plane. The terms "medial" and "lateral" relate to this plane.
- *Coronal (frontal) plane*—the plane dividing the body into equal or unequal front (anterior) and back (posterior) parts.
- *Transverse (cross sectional) plane*—the horizontal plane that divides the body into unequal upper (cranial) and lower (caudal) parts.

Several movements take place within each of these planes and it is important to understand these actions. A few of these movements are summarized in the following text with respect to the given planes:

TABLE 5.1
Common Body Measures and Their Applications

Dimensions	Applications
1. *Stature* The vertical distance from the floor to the top of the head, when standing	A main measure for comparing population samples. Reference for the minimal height of overhead obstructions. Add height for more clearance, hat, shoes, stride
2. *Eye height, standing* The vertical distance from the floor to the outer comer of the right eye, when standing	Origin of the visual field of a standing person. Reference for the location of visual obstructions and of targets such as displays; consider slump and motion
3. *Shoulder height (acromion), standing* The vertical distance from the floor to the tip (acromion) of the shoulder, when standing [2]	Starting point for arm length measurements; near the center of rotation of the upper arm. Reference point for hand reaches; consider slump and motion
4. *Elbow height, standing* The vertical distance from the floor to the lowest point of the right elbow, when standing, with the elbow flexed at 90°	Reference for height and distance of the work area of the hand and the location of controls and fixtures; consider slump and motion
5. *Hip height (trochanter), standing* The vertical distance from the floor to the trochanter landmark on the upper side of the right thigh, when standing	Traditional anthropometric measure, indicator of leg length and the height of the hip joint. Used for comparing population samples
6. *Knuckle height, standing* The vertical distance from the floor to the knuckle (metacarpal bone) of the middle finger of the right hand, when standing	Reference for low locations of controls, handles, and handrails; consider slump and motion of the standing person
7. *Fingertip height, standing* The vertical distance from the floor to the tip of the extended index finger of the right hand, when standing	Reference for the lowest location of controls, handles, and handrails; consider slump and motion of the standing person
8. *Sitting height* The vertical distance from the sitting surface to the top of the head, when sitting	Reference for the minimal height of overhead obstructions. Add height for more clearance, hat, trunk motion of the seated person
9. *Sitting eye height* The vertical distance from the sitting surface to the outer comer of the right eye, when sitting	Origin of the visual field of a seated person. Reference point for the location of visual obstructions and of targets such as displays; consider slump and motion
10. *Sitting shoulder height (acromion)* The vertical distance from the sitting surface to the tip (acromion) of the shoulder, when sitting	Starting point for arm length measurements; near the center of rotation of the upper arm. Reference for hand reaches; consider slump and motion
11. *Sitting elbow height* The vertical distance from the sitting surface to the lowest point of the right elbow, when sitting, with the elbow flexed at 90°	Reference for the height of an armrest, of the work area of the hand, and of keyboard and controls; consider slump and motion of the seated person

(continued)

TABLE 5.1 (continued)
Common Body Measures and Their Applications

Dimensions	Applications
12 *Sitting thigh height (clearance)* The vertical distance from the sitting surface to the highest point on the top of the horizontal right thigh, with the knee flexed at 90°	Reference for the minimal clearance needed between seat pan and the underside of a structure, such as a table or desk; add clearance for clothing and motions
13. *Sitting knee height* The vertical distance from the floor to the top of the right kneecap, when sitting, with the knees flexed at 90°	Traditional anthropometric measure for lower leg length. Reference for the minimal clearance needed below the underside of a structure, such as a table or desk; add height for shoe
14. *Sitting popliteal height* The vertical distance from the floor to the underside of the thigh directly behind the right knee; when sitting, with the knees flexed at 90°	Reference for the height of a seat; add height for shoe
15. *Shoulder-elbow length* The vertical distance from the underside of the right elbow to the right acromion, with the elbow flexed at 90° and the upper arm hanging vertically	Traditional anthropometric measure for comparing population samples
16. *Elbow-fingertip length* The distance from the back of the right elbow to the tip of the extended middle finger, with the elbow flexed at 90°	Traditional anthropometric measure. Reference for fingertip reach when moving the forearm in the elbow
17. *Overhead grip reach, sitting* The vertical distance from the sitting surface to the center of a cylindrical rod firmly held in the palm of the right hand	Reference for the height of overhead controls operated by a seated person. Consider ease of motion, reach, and finger/hand/arm strength
18. *Overhead grip reach, standing* The vertical distance from the standing surface to the center of a cylindrical rod firmly held in the palm of the right hand	Reference for the height of overhead controls operated by a standing person. Add shoe height. Consider ease of motion, reach, and finger/hand/arm strength
19. *Forward grip reach* The horizontal distance from the back of the right shoulder blade to the center of a cylindrical rod firmly held in the palm of the right hand	Reference for forward reach distance. Consider ease of motion, reach, and finger/hand/arm strength
20. *Arm length, vertical* The vertical distance from the tip of the right middle finger to the right acromion, with the arm hanging vertically	A traditional measure for comparing population samples. Reference for the location of controls very low on the side of the operator. Consider ease of motion, reach, and finger/hand/arm strength
21. *Downward grip reach* The vertical distance from the right acromion to the center of a cylindrical rod firmly held in the palm of the right hand, with the arm hanging vertically	Reference for the location of controls low on the side of the operator. Consider ease of motion, reach, and finger/hand/arm strength

TABLE 5.1 (continued)
Common Body Measures and Their Applications

Dimensions	Applications
22. *Chest depth* The horizontal distance from the back to the right nipple	A traditional measure for comparing population samples. Reference for the clearance between seat backrest and the location of obstructions in front of the trunk
23 *Abdominal depth, sitting* The horizontal distance from the back to the most protruding point on the abdomen	A traditional measure for comparing population samples. Reference for the clearance between seat backrest and the location of obstructions in front of the trunk
24. *Buttock-knee depth, sitting* The horizontal distance from the back of the buttocks to the most protruding point on the right knee, when sitting with the knees flexed at 90°	Reference for the clearance between seat backrest and the location of obstructions in front of the knees
25. *Buttock-popliteal depth, sitting* The horizontal distance from the back of the buttocks to back of the right knee just below the thigh, when sitting with the knees flexed at 90°	Reference for the depth of a seat
26. *Shoulder breadth (biacromial)* The distance between the right and left acromion	A traditional measure for comparing population samples. Indicator of the distance between the centers of rotation of the two upper arms
27. *Shoulder breadth (bideltoid)* The maximal horizontal breadth across the shoulders between the lateral margins of the right and left deltoid muscles	Reference for the lateral clearance required at shoulder level. Add space for ease of motion and tool use
28. *Hip breadth, sitting* The maximal horizontal breadth across the hips or thighs, whatever is greater, when sitting	Reference for seat width. Add space for clothing and ease of motion
29. *Span* The distance between the tips of the middle fingers of the horizontally outstretched arms and hands	A traditional measure for comparing population samples. Reference for sideway reach
30. *Elbow span* The distance between the tips of the elbows of the horizontally outstretched upper arms when the elbows are flexed so that the fingertips of the hands meet in front of the trunk	Reference for the lateral space needed at upper body level for ease of motion and tool use
31. *Head length* The distance from the glabella (between the browridges) to the most rearward protrusion (the occiput) on the back, in the middle of the skull	A traditional measure for comparing population samples. Reference for headgear size

(continued)

TABLE 5.1 (continued)
Common Body Measures and Their Applications

Dimensions	Applications
32. *Head breadth* The maximal horizontal breadth of the head above the attachment of the ears	A traditional measure for comparing population samples. Reference for headgear size
33. *Hand length* The length of the right hand between the crease of the wrist and the tip of the middle finger, with the hand flat	A traditional measure for comparing population samples. Reference for hand tool and gear size. Consider manipulations, gloves, tool use
34 *Hand breadth* The breadth of the right hand across the knuckles of the four fingers	A traditional measure for comparing population samples. Reference for hand tool and gear size, and for the opening through which a hand may fit. Consider manipulations, gloves, tool use
35. *Foot length* The maximal length of the right foot, when standing	A traditional measure for comparing population samples. Reference for shoe and pedal size
36. *Foot breadth* The maximal breadth of the right foot, at right angle to the long axis of the foot, when standing	A traditional measure for comparing population samples. Reference for shoe size, spacing of pedals
37. *Weight (kg)* Nude body weight taken to the nearest tenth of a kilogram	A traditional measure for comparing population samples. Reference for body size, clothing, strength, health, etc. Add weight for clothing and equipment worn on the body

Source: Adapted from Kroemer, K.H.E., *The Occupational Ergonomics Handbook*, Taylor & Francis, Boca Raton, FL, 2006.

Note: Descriptions of dimensions from Gordon et al. (1989) [26] with their reference numbers in brackets.

- Sagittal plane
  - *Flexion*—angle between two body segments is decreased.
  - *Extension*—angle between two body segments is increased.
  - *Dorsiflexion*—bringing the top of the foot toward the shin.
  - *Plantarflexion*—movement of the top of the foot away from the shin.
- Coronal plane
  - *Adduction*—movement toward the midline of the body.
  - *Abduction*—movement away from the midline of the body.
  - *Inversion*—lifting the medial border of the foot.
  - *Eversion*—lifting the lateral border of the foot away from the medial plane.
  - *Elevation*—moving to a superior position at the scapula.
  - *Depression*—moving to an inferior position at the scapula.
- Transverse plane
  - *Rotation*—external or internal turning on the vertical axis of a bone.

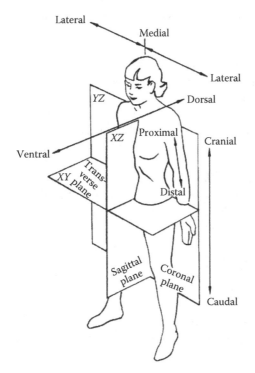

FIGURE 5.3 Anatomical landmarks used in measurement. (From NASA, Man-Systems Integration Standards, *Anthropometry and Biomechanics*, Volume I, Section 3, 2000.)

- *Pronation*—rotation of the hands and forearms medially from the elbow.
- *Supination*—rotation of the hands and forearms laterally from the elbow.

5.4.3 ANTHROPOMETRIC MEASUREMENTS

Measurements of anthropometric data are typically physical in nature, for example, height, weight, volume, density, range of motion, strength capabilities, and the distance between two landmarks on the body. However, sensory abilities may also be measured, such as hearing ability, sight, and the ability to sense touch may be considered in assessing a population. Specific terminology and measuring conventions in anthropometrics have been described by Garrett and Kennedy (1971); Gordon et al. (1989); Hertzberg (1968); Kroemer et al. (1990); Lohman et al. (1988); NASA/ Webb (1978); Pheasant (1986); Roebuck (1993); and Roebuck et al. (1975). These publications provide detailed information about traditional measurement procedures and techniques.

5.4.3.1 Stature Measurements

When taking stature measurements, it is essential that the subject takes on one of four traditional positions detailed in the following text. Regardless of the position in which the subject is measured, consistency of this position within the sample group for which measurements are taken is critical.

- Standing upright naturally
  - Slumping effect is included in measurements.
- Standing upright erect
  - Measurements can have a 2 cm difference when the subject standing either stretches to a fully erect position or just stands upright naturally.
- Standing against a wall with shoulder blades, buttocks, and back of the head touching the wall
  - Extending a book or straight edge from the top of the head to the wall, making a mark on the wall, and then measuring the distance to the floor helps ensure the subject is fully upright and provides a mechanism for consistency between subjects.
- Lying supine
  - Provides the tallest measurement as gravity will not compress the spine.
  - Taking measurements in the morning also provides the tallest measurement.

5.4.3.2   Seated Measurements

When taking measurements while the subject is seated, the following steps need to be considered:

- The seat and floor need to be parallel and horizontal.
- The knees need to be bent at a 90° angle.
- Thighs need to be placed in a horizontal position and the lower legs need to be placed in a vertical position.
- The feet have to be horizontally flat on the ground.

5.4.3.3   Measurements of Physical Properties of Body Segments

In ergonomic or biomechanical analysis when a person is performing a task, the human body is simplified to a system of mechanical links, each of known physical size and form.

Anthropometry defines these sizes and forms and seeks to determine various relevant properties such as length, volume, weight, location of the center of mass, and inertial properties. Each of these properties can be used in anthropometric design and biomechanical analysis. These values can also be estimated using body weight and statue. Figure 5.4a and 5.4b provide estimated segment length for males and females as a function of height (h).

The lengths of these segments can be estimated as a function of total stature.

5.4.4   Static and Dynamic Dimensions

Anthropometry can be divided into two sections: physical anthropometry, which consists of the basic dimensions of the human body in both standing and sitting positions, and functional anthropometry, which deals with task-oriented movements. Both physical and functional anthropometry can be regarded in either a static or dynamic manner. Static analysis suggests that only the measurements of the body segment lengths in fixed positions need to be considered in the design of workplaces. This is not always a valid assumption. Static dimensions are related to and vary with other factors, such as age, gender, ethnicity, occupation, and percentile within a specific population group and historical period (i.e., diet and living conditions). Stature changes as a result of aging, gravity, and approximate changes in stature with age are listed in Table 5.2 (Kroemer, 1994).

Body size of the 40-year-old American male for year 2000 in one gravity conditions

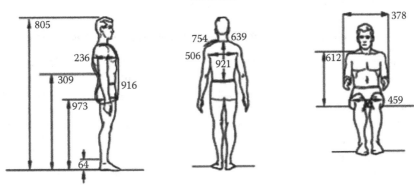

Dimension	5th percentile	50th percentile	95th percentile
Stature	169.7 (66.8)	179.9 (70.8)	190 1 (74.8)
Wrist height			
Ankle height	12.0 (4.7)	13.9 (5.5)	15.8 (6.2)
Elbow height			
Bust depth	21.8 (8.6)	25.0 (9.8)	28.2 (11.1)
Vertical trunk circumference	158.7 (62.5)	170.7 (67.2)	182.6 (71.9)
Midshoulder height, sitting	60.8 (23.9)	65.4 (25.7)	70.0 (27.5)
Hip breadth, sitting	34.6 (13.6)	38.4 (15.1)	42.3 (16.6)
Waist back	43.7 (17.2)	47.6 (18.8)	51.6 (20.3)
Interscye	32.9 (13.0)	39.2 (15.4)	45.4 (17.9)
Neck circumference	35.5 (14.0)	38.7 (15.2)	41.9 (16.5)
Shoulder length	14.8 (5.8)	16.9 (6.7)	19.0 (7.5)
Forearm-forearm breadth	48.8 (19.2)	55.1 (21.7)	61.5 (24.2)

Values in cm with inches in parentheses

FIGURE 5.4   (a) Anthropometric Dimensional Data for American Male, (b) Anthropometric Dimensional Data for American Female

Source: Johnson Space Center, NASA: msis.jsc.nasa.gov. Accessed August 15, 2011

(*continued*)

Body size of the 40-year-old Japanese female for year 2000 in one gravity conditions

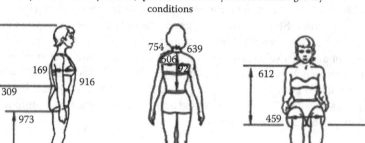

Dimension	5th percentile	50th percentile	95th percentile
Stature	148.9 (58.6)	157.0 (61.8)	165.1 (65.0)
Wrist height	70.8 (27.9)	76.6 (30.2)	82.4 (32.4)
Ankle height	5.2 (2.0)	6.1 (2.4)	7.0 (2.8)
Elbow height	92.8 (38.5)	98.4 (38.8)	104.1 (41.0)
Bust depth	17.4 (6.8)	20.5 (8.1)	23.6 (9.3)
Vertical trunk circumference	136.9 (53.9)	146.0 (57.5)	155.2 (61.1)
Midshoulder height, sitting			
Hip breadth, sitting	30.4 (12.0)	33.7 (13.3)	37.0 (14.6)
Waist back	35.2 (13.9)	38.1 (15.0)	41.0 (16.1)
Interscye	32.4 (12.8)	35.7 (14.1)	39.0 (15.4)
Neck circumference	34.5 (13.6)	37.1 (14.5)	39.7 (15.6)
Shoulder length	11.3 (4.4)	13.1 (5.1)	14.8 (5.8)

Values in cm with inches in parentheses

FIGURE 5.4 (continued)

On the other hand, dynamic analysis implies that the acceptability of a workplace design has to be evaluated with respect to the needed movements of the body from one position to another. Also, dynamic analysis takes into account clearance and reaching considerations. An example of important dynamic data that is frequently used in the design of workplaces is the range of joint mobility. Dynamic data is more costly and time-consuming to measure, which sometimes constrains occupational environments to using static measurement workplace designs.

In dynamic anthropometrics joint range of motion is an essential element in design. Figure 5.5 provide joint movement ranges for males and females for common actions.

5.4.4.1 Variability

As anthropometric data is collected, it is common to experience variability; however, if subjects are from the same population, consistency with anthropologic norms should be seen in the data as a normal distribution is expected. In the cases

TABLE 5.2
Approximate Changes in Stature with Age

| Age (Years) | Change (cm) | |
	Females	Males
1–5[a]	+36	+36
5–10	+27	+25
10–15	+22	+30
15–20	+7	+11
20–35[b]	−1	0
35–40	−1	0
40–50	−1	−1
50–60	−1	−1
60–70	−1	−1
70–80	−1	−1
80–90	−1	−1

Source: Kroemer et al., 1994.
[a] Average stature at age 1: females 72 cm, males 74 cm.
[b] Average maximal stature: females 163 cm, males 176 cm.

of variability, it is important to understand and to explain variation. Anthropometric data can contain three types of variability:

- Intraindividual variability
  - Changes in the individual over time due to age, health, diet
  - Traced through longitudinal studies
- Interindividual variability
  - Natural variation from individual to individual
  - Measured by a cross-section study (i.e., different individuals are measured at the same moment of time)
- Secular variability
  - Difference between generations
  - Evidence that people are larger than individuals 100 years ago

In some cases, the source of the variability may be because participants do not belong to the population of interest. Measurement variability can also result from inconsistent measuring techniques or equipment. For that reason, it is crucial to take multiple readings (at least three) of the same measurements. Additionally, when calculating descriptive statistics for evaluating the data, measures of central tendency, variation, and outliers should be identified.

5.4.5 Anthropometric Tools

Classical measurement techniques, some of which are rudimentary, are still widely used in anthropometric measurement calculations.

Figure	Joint movement (note b)	Range of motion (degrees)			
		Males (note a)		Female (note a)	
		5th percentile	95th percentile	5th percentile	95th percentile
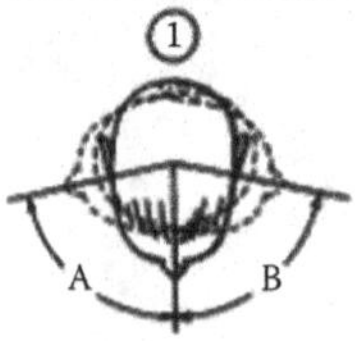 Neck rotation right (A) left (B)	Neck, rotation right (A)	73.3	99.6	74.9	108.8
	Neck, rotation left (B)	74.3	99.1	72.2	109.0
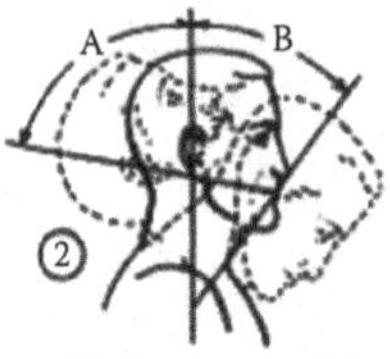 Neck extension (A) Flexion (B)	Neck, flexion (B)	34.5	71.0	46.0	84.4
	Neck, extension (A)	65.4	103.0	4.9	103.0
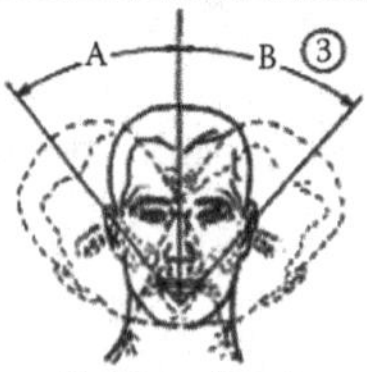 Neck lateral bend right (A) left (B)	Neck, lateral bend right (A)	34.9	63.5	37.0	63.2
	Neck, lateral bend left (B)	35.5	63.5	29.1	77.2
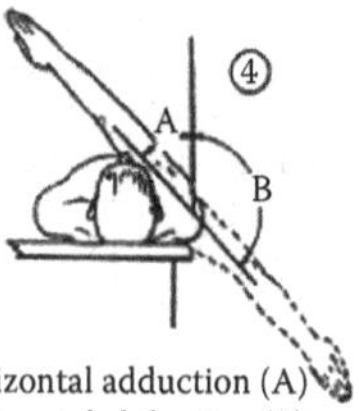 Horizontal adduction (A) Horizontal abduction (B)	Shoulder, abduction	173.2	188.7	172.6	192.9
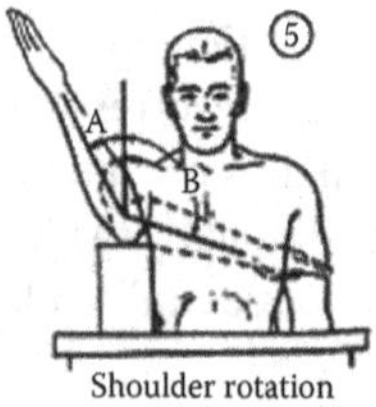 Shoulder rotation lateral (A) medial (B)	Shoulder, rotation lateral (A)	46.3	96.7	53.8	85.8
	Shoulder, rotation medial (B)	90.5	126.6	95.8	130.9

FIGURE 5.5   Joint Movement Ranges for Males and Females
Source: Johnson Space Center, NASA: msis.jsc.nasa.gov. Accessed August 15, 2011.

Figure	Joint movement (note b)	Range of motion (degrees)			
		Males (note a)		Female (note a)	
		5th percentile	95th percentile	5th percentile	95th percentile
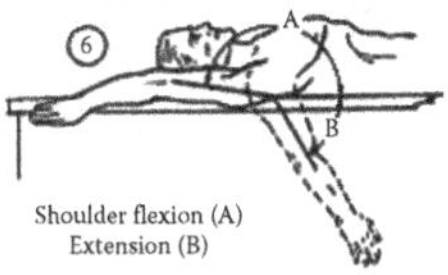 Shoulder flexion (A) Extension (B)	Shoulder, flexion (A)	164.4	210.9	152.0	217.0
	Shoulder, extension (B)	39.6	83.3	33.7	87.9
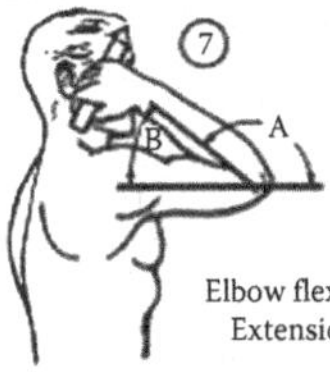 Elbow flexion (A) Extension (B)	Elbow, flexion (A)	140.5	159.0	144.9	165.9
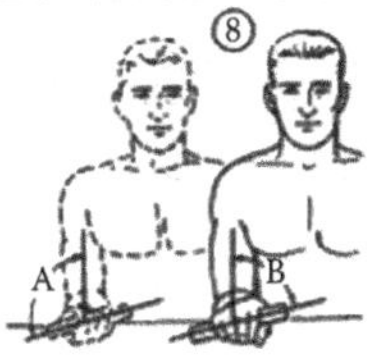 Forearm supination (A) Pronation (B)	Forearm, pronation (B)	78.2	116.1	82.3	118.9
	Forearm, supination (A)	83.4	125.8	90.4	139.5
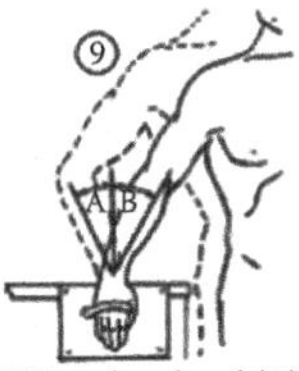 Wrist ulnar bend (A) Radial bend (B)	Wrist, radial bend (B)	16.9	36.7	16.1	36.1
	Wrist, ulnar bend (A)	18.6	47.9	21.5	43.0
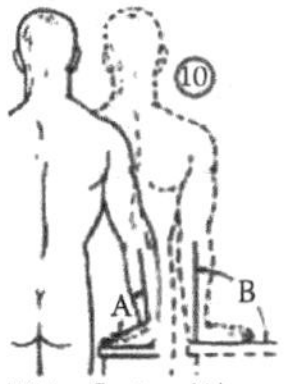 Wrist flexion (A) Extension (B)	Wrist, flexion (A)	61.5	94.8	68.3	98.1
	Wrist, extension (B)	40.1	78.0	42.3	74.7
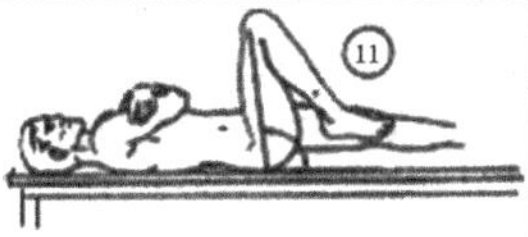 Hip flexion	Hip, flexion	116.5	148.0	118.5	145.0

FIGURE 5.5 (continued)

Figure	Joint movement (note b)	Range of motion (degrees)			
		Males (note a)		Female (note a)	
		5th percentile	95th percentile	5th percentile	95th percentile
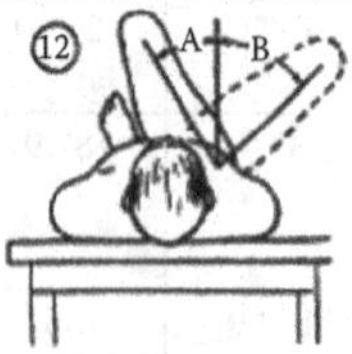 ⑫ Hip adduction (A) Abduction (B)	Hip, abduction (B)	26.8	53.5	27.2	55.9
 ⑬ Knee flexion, prone	Knee, flexion	118.4	145.6	125.2	145.2
 ⑭ Ankle plantar extension (A) Dorsi flexion	Ankle, plantar extension (A)	36.1	79.6	44.2	91.1
	Ankle, dorsi flexion (B)	8.1	19.9	6.9	17.4

Notes:
a. Data was taken 1979 and 1980 at NASA-JSC by Dr. William Thornton and John Jackson. The study was made using 192 males (mean age 33) 22 females (mean age 30) astronaut candidates.
b. Limb range is average of right and left limb movement.

FIGURE 5.5 (continued)

Anthropometric instruments are used to obtain lengths, widths, breadths, and joint angles of the human anatomy. Some of these tools include high-tech scanners, photometric equipment, and hand tools, such as anthropometers, calipers, and measuring tapes. An anthropometer is a 2 m long, graduated rod that has a sliding edge at a right angle.

- Spreading calipers have two curved branches that are joined in a hinge, and a scale is used to measure the distance between the tips. Short measurements, such as finger length or finger thickness, are calculated by using small sliding calipers.
- Special calipers are employed to measure the thickness of skinfolds, and cones are used to measure the diameter around which a finger can close. Another instrument, which is used to measure the external diameter of a finger, is a tool that has circular holes of different sizes drilled in a thin plate.
- Circumferences and curvature measurements are obtained by using tapes.

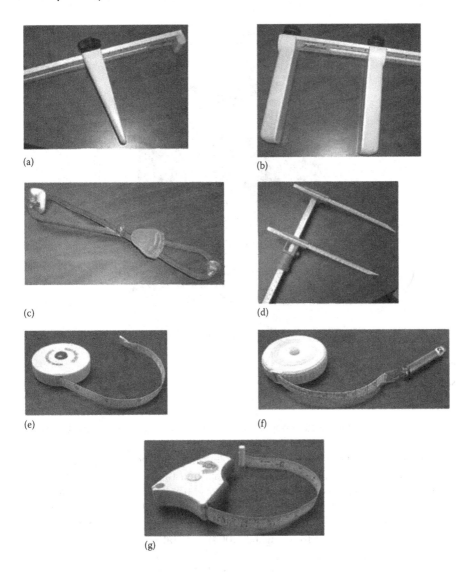

(a)

(b)

(c)

(d)

(e)

(f)

(g)

FIGURE 5.6 Manual anthropometric tools.

Example of these manual anthropometric tools are shown in Figure 5.6.

The majority of the conventional tools used for anthropometric measurements are applied by the measurer's hand to the body of the subject. Hand measurement instruments also have several shortcomings. Some of these tools are not practical for use in taking measurements on certain parts of the body, such as the eyes. They are also slow, cumbersome, and can be prone to human error. It is important to remember that anthropometric data always needs to be handled cautiously, and any data that was not gathered during the test session cannot be inferred and must remain unknown, until such time that additional measurements can be made. Also, because anthropometric

FIGURE 5.7   Air force.
Source: Air Force Research Laboratory, Wright Patterson Air Force Base.

guidelines represent averages of populations, this information should simply be used as a guideline, not as an absolute in design.

New measurement techniques, such as photographs, can record three-dimensional characteristics of the human body, allowing an almost infinite number of measurements to be taken. However, photographs can also have disadvantages in anthropometrics. The human body is portrayed in two dimensions and thus scale and parallax errors can occur. However, this is changing with enhanced photo-anthropometric technology (Davis et al., 2010; Hungetal, 2004). Several techniques have been proposed in order to acquire three-dimensional anthropometric data.

The Air Force Research Laboratory at Wright Patterson Air Force Base uses 3D anthropometric full-body scanner in human factors research. This 3D automatic full body scanner works while a test subject model poses in the standing scan position in the scanner (Figure 5.7). Researchers place 72 white stickers at key anatomical sites on the body of the test subject. The subject is fully scanned and the site stickers allow researchers to take additional measurments from scanned data, using the stickers as reference points. An example is the laser, which is used to measure the shape of irregular bodies. Laser-based anthropometry provides fast, accurate, and multiple three-dimensional readings. However, this technique can be very costly. A recently developed automated measuring tool is provided in the following section.

5.4.5.1 Automated Anthropometric Measurement Tools

As new technologies emerge, anthropometric measurement tools are moving toward automation. The use of automated technologies has led to the development of anthropometric measuring tools that increase precision and efficiency while reducing error. Once such product is the anthropometric measurement system (AMS) created by Ergonomic Technologies (Ergotech, 2011). The AMS is an electromechanical body dimension-measuring instrument that measures critical body dimensions and electronically transfers it to a personal computer-based software system, where clothing item sizes are allocated according to the body dimensions of the individual (Ergonomic Technologies, 2010). The main purpose of AMS is to ensure proper issuance of clothing to defense personnel.

Human Systems Integration Information Analysis Center (HSIIAC)—formerly Crew System Ergonomics Information Analysis Center (CSEIAC)—provides Human Systems Integration (HSI) information and analysis services in support of research, design, development, test, and evaluation of human-operated systems (HSIIAC website, 2011). In a report published for the CSEIAC, Moroney references another system that is in development, the automated anthropometric data measurement system (AADMS). Both of these systems are examples of the automated anthropometric measuring tools currently in use or development and show the move from manual to automated measuring tools for anthropometric measurement purposes.

5.4.6 ANTHROPOMETRIC DESIGN AIDS AND DATA SOURCES

Anthropometric design aids can include data tables, templates, mannequins, two- and three-dimensional surface models, and computer-generated models. All of these aids rely on historic data on human body dimensions. These anthropometric data sources are available to serve as baseline information for comparison or to be used directly in design. The compatibility of the population of interest and the particular data source must be considered in selecting an appropriate database. Examples of anthropometric databases are provided in the following section; however, this is not an attempt to minimize the value or significance of other data sources. Each source is significant and appropriate in specific applications; however, no endorsement is made of any resource.

5.4.6.1 Civilian American and European Surface Anthropometry Resource

The Civilian American and European Surface Anthropometry Resource (CAESAR) began as a partnership between government and industry to collect an extensive sampling of consumer body measurements for comparison, contrast, and application. The project collected data on 2400 American and Canadian subjects and 2000 European civilians and a created a database to store this information (SAE International, 2010).

The CAESAR database contains anthropometric variability of men and women, ages 18–65. A broad range of representatives were solicited to ensure samples for various weights, ethnic groups, gender, geographic regions, and socioeconomic status. The study was conducted from April 1998 to early 2000 and includes three scans per person in a standing pose, full-coverage pose, and relaxed seating pose. Data collection methods were standardized and documented so that the database can be consistently expanded and updated (Ashdown et al., 2004).

The CAESAR product line was designed to provide reliable and current measurements of the human body for the current generations. This product line was developed as a result of a comprehensive research project that brought together representatives from numerous industries including apparel, aerospace, and automotive. The CAESAR product line is available to researchers, students, academics, and designers throughout the world.

5.4.6.2 U.S. Army Anthropometry Survey

The 1988 U.S. Army Anthropometry survey is one of the most widely used anthropometry databases because of the large number of measurements and the rigorous methodology used to collect the data. This data is most useful for analyzing relationships between anthropometric variables, since the sample population (U.S. Army as of 1988) is usually not representative of any particular target population. The data can be readily applied to groups of the population that tend to have characteristics similar to those of U.S. Army soldiers (i.e., comparable fitness levels, age, weight, etc.).

5.4.6.3 AnthroKids

Although many people believe that children are just little adults, this is not the case. Children have unique anthropometric characteristics and must be designed for accordingly. An example of the differences in anthropometrics for children compared to adults is the higher center of gravity due to the fact that the head is a larger proportion of total body weight with small children. The National Institute of Standards and Technology commissioned a collection of anthropometric data on children. The first study was performed in 1975 and followed up by another study in 1977. This study collected anthropometric data on children with an emphasis on measurements that could be used to support designers in the development of products for children, with a particular focus on safety concerns. (Ressler S., 2011).

5.4.6.4 International Anthropometric Resource

The global community of anthropometry has accessible resources to obtain anthropometric data. One such resource international ergonomists can access is the World Engineering Anthropometry Resource (WEAR). WEAR is an international collaborative effort to create a worldwide resource of anthropometric data for a wide variety of engineering applications (National Institute of Standards and Technology, 2007). Also, the U.S. National Aeronautics and Space Administration (NASA, 1978) published a reference publication with measures of 306 different body dimensions from 91 different populations around the world.

5.4.6.5 The International Journal of Clothing Science and Technology

The International Journal of Clothing Science and Technology boasts an article in the third issue in Volume 18 of a Croatian anthropometric system. This article includes information on the first systematic anthropometric measurement in all Croatian counties with an objective to determine a proposal of the new size system of clothing and footwear (Darko et al., 2006). An organization that provides anthropometric data from a variety of countries is Humanics Ergonomics, a private consulting firm. This firm has collected a broad data set that includes anthropometric data from children to disabled, for many countries around the world (Humanics, 2010).

5.4.7 Procedure for Design Using Anthropometric Data

To obtain consistent and reliable measures, certain methods should be followed when designing an anthropometric study and in the collection of data. The following list provides a guideline for conducting anthropometric studies:

- Characterize the user population.
  - Identify the type of anthropometric data available, and whether or not the existing anthropometric data is valid for the present population.
  - If reliable anthropometric data exists, use the available data to support design decisions.
  - If there is no valid data, perform an analysis and create a database by obtaining measures of a sample population that is representative of the existing workforce.
- Determine a sample size to be used for collecting data from the population of interest.
- Determine the percentile range to be accommodated in the design.
  - If the workforce or environment is dominated by either men or women, it may be reasonable to give design consideration for the predominant gender, for example, by using the 5th to 95th percentile male or 5th to 95th percentile female measures.
  - It may be an issue of equality to provide accessibility for the other gender group and in these situations, design for the 5th percentile female to the 95th percentile male population.
  - Determine population size and characteristics.
- Design workspaces and tasks such that the smallest person can reach all items and the largest person can fit into all spaces.
  - Determine reach dimensions (5th percentile).
  - Determine clearance dimensions (95th percentile).
- Identify relevant anthropometric measures.
- Collect anthropometric data.
- Obtain relevant statistics.
- Compare data to relevant databases.
- Determine applicable ranges and measures to be used in the study outcomes.

Statistical analysis is vital in the analysis and use of anthropometric calculations. Measures of central tendency, such as the mean, median, and mode, as well as measures of variability including the range, standard deviation, standard error, skewness, and peakedness should be computed. Measures of relationship, such as the correlation coefficient and regression analysis, should be determined.

Confidence intervals (CIs) that show the varying degrees of confidence are significant in anthropometric studies. Generally, CIs are established that attempt to accommodate 90%–95% of the population. CIs are based on normal distributions and on the historic, statistical empirical rule. The empirical rule states that for a normally distributed data set, approximately

- 68.76% of the data set will be within 1 standard deviation.
- 95.65% of the data set will be within 2 standard deviations.
- 99.73% of the data set will be within 3 standard deviations.

These intervals can be used to construct approximate ranges for a population; however, to construct a CI, it is necessary to know the mean and standard deviation for a data set and define the level of desired confidence in the interval. Assuming normality a random sample and a data set containing at least thirty data points, the following applies:

$$95\% \ CI = \mu \pm 1.96\sigma \tag{5.1}$$

$$90\% \ CI = \mu \pm 1.65\sigma \tag{5.2}$$

Where
µ represents population mean
σ represents population standard deviation

For smaller samples, that are normally distributed the t-distribution may be used however, the CI's will have a larger range due to a reduced sample size. In this case, the CI formulas become

$$95\% \ CI = \mu \pm t\sigma \tag{5.3}$$

$$90\% \ CI = \mu \pm t\sigma \tag{5.4}$$

Where
µ represents population mean
σ represents population standard deviation
t –represents the value for the given sample size and the associated degrees of freedom

Statistical analysis should also examine the variability in a collected set of anthropometric data. Calculating values such as the coefficient of variation can reveal variability within a data set and provide insight regarding the normality of the distribution.

5.4.8 ANTHROPOMETRICS AND BIOMECHANICS

Biomechanics is a field related to anthropometry that studies the human body in terms of a mechanical system. In order to understand how to model the human body,

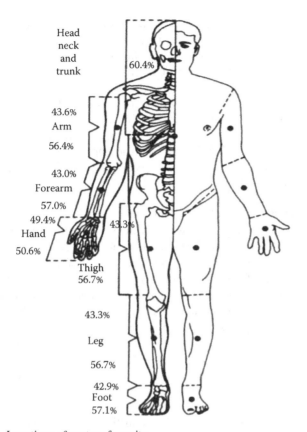

FIGURE 5.8   Locations of center of gravity.

a number of parameters must be understood including link length measurements (lengths, widths, and circumferences), volume and weight, mass center locations, and inertial property measurements. Assuming that the body is connected at easily identifiable points or landmarks, it is necessary to define these anatomical landmarks to measure the distances. Commonly used anatomical landmarks include the segment joints. Body segment link length measurements are obtained with linear or angular measurement devices. This is useful model to estimate the location of the center of gravity for humans.

A variety of anthropometric data is available on segment weights, and this information is used to identify the center of gravity locations for these segments. The location of the center of gravity is shown in Figure 5.8 and is located at the given percentage below the proximal end of the segment. Mass center is the location from which a segment can be suspended and have complete balance. Cadaver data has been used extensively to obtain mass center locations. Using this data, the approximate location of mass centers can be taken as a percentage of the link length. Generally, the subject lays horizontally on a force platform and the center of gravity for a particular limb can be calculated. Also, immersion methods are used to determine the center of gravity. For example, a limb is immersed into the water up to proximal end, where the water goes to overflow the tank. The limb is removed when one half of the water returns to the tank because at this point the water level should bisect the limb's mass center. Upon determining the center of gravity for a link, the location to the center of gravity is defined as the distance from the proximal to distal ends of the link.

To understand where to take measurements, it is important to know where the joint centers or rotations lie for the various joint. The measurements are then made from one joint to another's center of rotation. The body is treated as a system of links in order to perform a biomechanical analysis. The lengths are estimated or measured anthropometrically, and then the assessments are made in biomechanical modeling such as length analysis.

5.4.8.1 Anthropometry in Application

Anthropometric guidelines have played a critical role in different types of organizations including the airline and clothing industries. The degree of anthropometric study and application in a design is a function of many factors, including but not limited to, the impact of poor anthropometric design, economics, user expectations and in some cases, risk of litigation. Studies have not demonstrated a direct safety impact associated with smaller seats in aircraft and thus it appears that anthropometric principles have been largely ignored in passenger airline seating. Additionally, introducing changes in seating dimensions has the potential to have a substantial economic impact, thus making this undesirable. The fact that the smaller and less comfortable seating has not been a primary differentiator in travelers' selection of an airline is even more motivation to delay the implementation of these principles and designs that consider larger passenger dimensions. Anthropometric measurements that continue to be of interest in the airline industry include limb length, trunk dimensions, popliteal height, hip dimensions, and more. Devices such as seat belt extenders are used for larger passengers as well as the emerging requirement for larger passengers to purchase two seats. However, additional applications of anthropometric data are clearly needed in the commercial airline industry.

Anthropometric guidelines have been significant in the clothing industry. For example, how is it possible that a 5′8″, 150 lb woman, a 5′6″, 135 lb woman, and a 5′9″, 125 lb woman can all wear a size 8. These fitting anomalies are the reality today for the clothing industry, where a comprehensive analysis of body shapes and sizes has not been conducted for decades. The SizeUSA survey was an anthropometric research project designed to address this need in the U.S. clothing industry. Researchers gathered U.S. sizing data with the use of a three-dimensional measurement system and body scanner, which fed the data into measurement extraction software. The body measurement system consisted of four strategically placed cameras that used white light to register more than 200,000 data points on the body. These points were reduced to 40,000 and become a point cloud of the body. The results are 200 accurate body measurements in less than 1 min. This project scanned over 10,000 subjects who were grouped into gender, six age groups, and four ethnicities. The survey filled 48 statistical categories that could be utilized in a variety of ways to support the needs of clothing manufacturers. A detailed listing of anthropometric data sources is contained at the end of this chapter (Tables 5.3 and 5.4).

TABLE 5.3
Anthropometric Data (cm)[a]

Measurement	Males 50th Percentile	Males ±1 SD	Females 50th Percentile	Females ±1 SD	Population Percentiles, 50/50 Males/Females 5th	50th	95th
Standing							
1. Forward functional reach							
a. Includes body depth at shoulder	82.6 (79.3)	4.8 (5.6)	74.1 (71.3)	3.9 (4.4)	69.1 (65.5)	77.9 (74.8)	88.8 (86.5)
b. Acromial process to functional pinch	63.8 (62.1)	4.3 (8.9)	62.5 (60.4)	3.4 (6.7)	57.5 (48.5)	65.0 (61.1)	74.5 (74.5)
c. Abdominal extension to functional pinch[b]	23.1	2.0	20.9	2.1	18.1	22.0	25.8
2. Abdominal extension depth							
3. Waist height	106.3 (104.8)	5.4 (6.3)	101.7 (98.5)	5.0 (5.5)	94.9 (91.0)	103.9 (101.4)	113.5 (113.0)
4. Tibial height	45.6	2.8	42.0	2.4	38.8	43.6	49.2
5. Knuckle height	75.5	4.1	71.0	4.0	65.7	73.2	80.9
6. Elbow height	110.5 (114.6)	4.5 (6.3)	102.6 (107.1)	4.8 (6.8)	96.4 (98.8)	106.7 (110.7)	116.3 (123.5)
7. Shoulder height	143.7 (146.4)	6.2 (7.8)	132.9 (135.3)	5.5 (6.6)	124.8 (126.6)	137.4 (140.4)	151.7 (156.4)
8. Eye height	164.4	6.1	151.4	5.6	144.2	157.7	172.3
9. Stature	174.5 (177.5)	6.6 (6.7)	162.1 (164.5)	6.0 (7.2)	154.4 (155.1)	168.0 (170.4)	183.0 (188.7)
10. Functional overhead reach	209.6	8.5	199.2	8.6	188.0	204.5	220.8
Seated							
11. Thigh clearance height	14.7	1.4	12.4	1.2	10.8	13.5	16.5
12. Elbow rest height	24.1	3.2	23.1	3.0	18.4	23.6	28.9
13. Midshoulder height	62.4	3.2	58.0	2.7	54.5	60.0	66.5
14. Eye height	78.7	3.6	73.7	3.1	69.7	76.0	83.3
15. Sitting height normal	86.6	3.8	81.8	4.0	76.6	84.2	91.6
16. Functional overhead reach	128.4	8.5	119.8	6.6	110.6	123.6	139.3
17. Knee height	54.0	2.7	51.0	2.6	47.5	52.5	57.7

(continued)

TABLE 5.3 (continued)
Anthropometric Data (cm)[a]

Measurement	Males		Females		Population Percentiles, 50/50 Males/Females		
	50th Percentile	±1 SD	50th Percentile	±1 SD	5th	50th	95th
18. Popliteal height	44.6	2.5	41.0	1.9	38.6	42.6	47.8
19. Leg length	105.1	4.8	100.7	4.3	94.7	102.8	111.4
20. Upper-leg length	59.4	2.8	57.4	2.6	53.7	58.4	63.3
21. Buttocks-to-popliteal length	49.8	2.5	48.0	3.2	43.8	49.0	53.6
22. Elbow-to-fist length	38.5 (37.1)	2.1 (3.0)	34.8 (32.9)	2.3 (3.1)	31.9 (28.9)	36.7 (35.0)	41.1 (41.0)
23. Upper-arm length	36.9 (37.0)	1.9 (2.5)	34.1 (33.8)	2.5 (2.1)	31.0 (28.9)	35.7 (35.0)	39.4 (41.0)
24. Shoulder breadth	45.4	1.9	39.0	2.1	36.3	42.3	47.8
25. Hip breadth	35.6	2.3	38.0	2.6	32.4	36.8	41.5
Foot							
26. Foot length	26.8	1.3	24.1	1.1	22.6	25.3	28.4
27. Foot breadth	10.0	0.6	8.9	0.5	8.2	9.4	10.8
Hand							
28. Hand thickness, metacarpal III	3.3	0.2	2.8	0.2	2.7	3.0	3.6
29. Hand length	19.0	1.0	18.4	1.0	17.0	18.7	20.4
30. Digit two length	7.5	0.7	6.9	0.8	5.8	7.2	8.5
31. Hand breadth	8.7	0.5	7.7	0.5	7.0	8.2	9.3
32. Digit one length	12.7	1.1	11.0	1.0	9.7	11.8	14.2
33. Breadth of digit one interphalangeal joint	2.3	0.1	1.9	0.1	1.8	2.1	2.5
34. Breadth of digit three interphalangeal joint	1.8	0.1	1.5	0.1	1.4	1.7	2.0

35. Grip breadth, inside diameter	4.9	0.6	4.3	0.3	3.8	4.5	5.7
36. Hand spread, digit one to digit two, first phalangeal joint	12.4	2.4	9.9	1.7	7.5	10.9	15.5
37. Hand spread, digit one to digit two, second phalangeal joint	10.5	1.7	8.1	1.7	5.9	9.3	12.7
Head							
38. Head breadth	15.3	0.6	14.5	0.6	13.8	14.9	16.0
39. Interpupillary breadth	6.1	0.4	5.8	0.4	5.2	6.0	6.7
40. Biocular breadth	9.2	0.5	9.0	0.5	8.3	9.1	10.0
Other measurements							
41. Flexion-extension, range of motion of wrist radians (57°/rad)	2.33	0.33	2.46	0.26	1.92	2.4	2.8
42. Ulnar-radial range of motion of wrist radians (57°/rad)	1.05	0.23	1.17	0.24	0.81	1.15	1.49
43. Weight, in kilograms	83.2	15.1	66.4	13.9	47.7	74.4	102.9

Source: Adapted from Champney, P.C. (1979); Muller-Borer, B. (1981); Eastman Kodak Company, *Ergonomics Design for People at Work*, Vol. 1, Van Nostrand Reinhold, New York, 1983; NASA RP 1024, *Anthropometric Source Book: Volume 1: Anthropometry for Designers Anthropology Staff/Webb Associates*, NASA, Washington, DC, pp. 7–78, 1978.

Note: The mean, or 50th percentile, values, plus or minus one standard deviation (SD) of the mean, are shown for 43 anthropometric variables (column 1). Variables 1 through 10 are standing heights, clearances, or reaches, and variables 11 through 25 are measurements for the subject seated. Data on American men (columns 2 and 3) and women (columns 4 and 5) are statistically combined to derive the 5th, 50th, and 95th percentile values for a 50/50 mix of these populations (columns 6 through 8). The data are taken primarily from military studies, where several thousand people were studied. The entries shown in parentheses are from industrial studies, where 50–100 women and 100–150 men were studied. The data in the footnotes "a and b" are from a study on 50 women and 100 men in the industry.

[a] These values should be adjusted for clothing and posture.

[b] Add the following for bending forward from hips or waist—Male: waist, 25 ± 7; hips, 42 ± 8. Female: waist, 20 ± 5; hips, 36 ± 9.

TABLE 5.4

International Anthropometry: Adults, Height, and Weight Averages (with Standard Deviations)

	Sample Size	Stature (mm)	Weight (kg)
Algeria			
Females (1990)	666	1576 (56)	61 (1)
Australia			
Females, 77 (8) years old	138	1521 (70)	61 (13)
Males, 76 (7) years old (2000)	33	1658 (79)	72 (11)
Brazil			
Males (1988)	3076	1699 (67)	NDA
China			
Females (Hong Kong)	69	1607 (54)	NDA
Females (Taiwan) (1994)	300	1582 (49)	51 (7)
Females (Taiwan) (2000)	About 600	1572 (53)	52 (7)
Males (Hong Kong) (2000)	286	1737 (49)	NDA
Males (Canton) (1990)	41	1720 (63)	60 (6)
Males (Taiwan) (2002)	About 600	1705 (59)	67 (9)
Egypt			
Females (1987)	4960	1606 (72)	63 (4)
France			
Female soldiers	328	1620	58
Male soldiers (1997)	687	1747	70
Females	5510	1625 (71)	62 (12)
Males (IFTH and Goncalves, personal communication, 2006)	3986	1756 (77)	77 (13)
Germany (east)			
Females	123	1608 (59)	NDA
Males (1986)	30	1715 (66)	NDA
India			
Females	251	1523 (66)	50 (10)
Males (1997)	710	1650 (70)	57 (11)
East-Central India male farm workers (2002)	300	1638 (56)	57 (7)
Central India male farm workers (1989)	39	1620 (50)	49 (6)
South India male workers (1992)	128	1607 (60)	57 (5)
East India male farm workers (1997)	134	1621 (58)	54 (67)
Indonesia			
Females	468	1516 (54)	NDA
Males (1985)	949	1613 (56)	NDA
Iran			
Female students	74	1597 (58)	56 (10)
Male students (1997)	105	1725 (58)	66 (10)
Ireland			
Males (1991)	164	1731 (58)	74 (9)

TABLE 5.4 (continued)
International Anthropometry: Adults, Height, and Weight Averages (with Standard Deviations)

	Sample Size	Stature (mm)	Weight (kg)
Italy			
Females (1991)	753	1610 (64)	58 (8)
Females (2002)	386	1611 (62)	58 (9)
Males (1991)	913	1733 (71)	75 (10)
Males (2002)	410	1736 (67)	73 (11)
Jamaica			
Females	123	1648	61
Males (1991)	30	1749	68
Japan			
Females	240	1584 (50)	54 (6)
Males (1990)	248	1688 (55)	66 (8)
Korea (South)			
Female workers (1989)	101	1580 (57)	54 (7)
Malaysia			
Females (1988)	32	1559 (66)	NDA
The Netherlands			
Females, 20–30 years old (1998)	68	1686 (66)	67 (10)
Females, 18–65 years old (2002)	691	1679 (75)	73 (16)
Males, 20–30 years old (1998)	55	1848 (80)	81 (14)
Males, 18–65 years old (2002)	564	1813 (90)	84 (16)
Russia			
Female herders (ethnic Asians)	246	1588 (55)	NDA
Female students (ethnic Russians)	207	1637 (57)	61 (8)
Female students (ethnic Uzbeks)	164	1578 (49)	56 (7)
Female factory workers (ethnic Russians)	205	1606 (53)	61 (8)
Female factory workers (ethnic Uzbeks)	301	1580 (54)	58 (9)
Male students (ethnic Russians)	166	1757 (56)	71 (9)
Male students (ethnic Uzbeks)	150	1700 (52)	65 (7)
Male factory workers (ethnic Russians)	192	1736 (61)	72 (10)
Male factory workers (ethnic mix)	150	1700 (59)	68 (8)
Male farm mechanics (ethnic Asians)	520	1704 (58)	64 (8)
Male coal miners (ethnic Russians)	150	1801 (61)	NDA
Male construction workers (ethnic Russians) (1999)	150	1707 (69)	NDA
Saudi Arabia			
Males (1985)	1440	1675 (61)	NDA
Singapore			
Females (1988)	46	1598 (58)	NDA
Males (pilot trainees) (1995)	832	1685 (53)	NDA
Sri Lanka			
Females	287	1523 (59)	774 (22)
Males (1991)	435	1639 (63)	833 (27)

(continued)

TABLE 5.4 (continued)
International Anthropometry: Adults, Height, and Weight Averages (with Standard Deviations)

	Sample Size	Stature (mm)	Weight (kg)
Sudan			
Males			
Villagers (1981)	37	1687 (63)	NDA
City dwellers (1982)	16	1704 (72)	NDA
City dwellers (1982)	48	1668	NDA
Soldiers(1981)	21	1735 (71)	NDA
Soldiers (1982)	104	1728	NDA
Thailand			
Females	250	1512 (48)	NDA
Females	711	1540 (50)	817 (27)
Males	250	1607 (20)	NDA
Males (1991)	1478	1654 (59)	872 (32)
Turkey			
Females			
Villagers	47	1567 (52)	792 (38)
City dwellers	53	1563 (55)	786 (05)
Male soldiers (1991)	5108	1702 (60)	888 (34)
United States			
Females	About 3800	1625	NDA
Males (2004)	About 3800	1762	NDA
Midwest workers/with shoes and light clothes			
Females	125	1637 (62)	NDA
Males (1993)	384	1778 (73)	NDA
U.S. male miners (1993)	105	1803 (65)	NDA
U.S. Army soldiers			
Females	2208	1629 (64)	852 (35)
Males (1989)	1774	1756 (67)	914 (36)
North American (Canada and United States)			
Females,18–26 years old	1255	1640 (73)	NDA
Males, 18–65 years old (2002)	1120	1778 (79)	NDA
Vietnamese, living in the United States			
Females	30	1559 (61)	NDA
Males (1993)	41	1646(60)	NDA

Source: Kumar, S. (Ed.), *Biomechanics in Ergonomics*, 2nd edn., CRC Press, New York, 2008.

5.5   SUMMARY

Anthropometrics enables us to properly design equipment, processes, and product items, including system interfaces, to accommodate the user. Accurate data on height, weight, limb, and body segment sizes are needed to properly design items ranging from clothing, furniture, automobiles, buses, and subway cars to space shuttles and space stations. The collection, interpretation, and utilization of anthropometric data will lead to enhanced ergonomics in occupational environments, leisure settings, and product design. Safety issues related to consumer products, particularly those designed for use by children, continue to surface, sometimes with tragic consequences. The need for continual collection of anthropometric data is reinforced by product liability lawsuits, the continuing trend toward obesity in numerous cultures and development of products for an international cross-section of consumers.

Case Study

OSHA Success with Ergonomics

International Truck and Engine Corporation

State: Ohio

Company: International Truck and Engine Corporation, Springfield, Ohio Assembly Plant

Industry: Motor vehicle and passenger car bodies—SIC Code: 3711

Employees: 2500

Success Brief: The redesign of workstation layouts, improvement of racking and storage, and reevaluation and replacement of certain tools led to a significant decrease in incident frequencies and lost time cases caused by musculoskeletal disorders.

The Problem

Four years before it launched a new series of trucks at its Springfield Assembly Plant, International Truck and Engine set goals to decrease or eliminate ergonomic risks and other potential problems associated with production, while at the same time increasing the facility's productivity and efficiency.

The Solution

Before the launch of the new product line, management, production employees and safety and health representatives worked together in teams to redesign the workspaces. The employees raised concerns during workshops involving cross-functional teams that included skilled trades employees, line supervisors, maintenance supervisors, safety and ergonomics representatives, upper management representatives, and production line employees. During the workshops, the team members learned about ergonomic risk factors, the importance of workplace organization and set up, and various safety procedures to be used in their workplace. As part of the redesign, the company:

> Installed lift and tilt tables so that employees could adjust the workstation height as needed, which reduced incidents of shoulder and back strains and sprains.
>
> Hung and balanced all tools overhead, raised air hoses off the floor, and replaced pistol-grip tools with inline tools, which reduced wrist and elbow injuries and eliminated trip and fall hazards.
>
> Redesigned flow-through racks to incorporate adjustable shelf heights so that employees of various heights could keep parts on racks within an employee's own personal "strike zone" for improved lifting.
>
> Altered the radiator assembly line to improve the employees' access to the radiator under assembly by installing variable height running boards and a "kick lever" to allow rotation of the radiator as needed by the assembly line employees.
>
> Standardized the containers, rack design, and stock positioning so that heavier items were carried the shortest distance and smaller parts were placed in standardized totes on rolling racks that allow for employee adjustment, depending on the employee's needs.

Source: Case Study reprinted as shown in OSHA website "Success Stories", accessed 2011; originally printed in *Occupational Health & Safety*, Volume 71, Issue 9, September 1, 2002. Updated, Subhash C. Vaidya, Navistar International Corporation, October 20.

EXERCISES

5.1 Explain the purpose for universal design and how this might impact the economics, dimensions, and processes of design from an anthropometric perspective.

5.2 In the absence of universal design, what ranges should the ergonomist promote design for?

5.3 In emergency situations, how should anthropometric data be used to guide design decisions?

5.4 Develop a process for collecting anthropometric data in an occupational ergonomic environment.

5.5 Contrast and compare the use of manual versus automatic anthropometric measurement tools.

5.6 Evaluate the use of goniometers for measuring body motions.

5.7 Describe how stature is measured and why it is relevant to occupational ergonomics.

5.8   Explain why the dimensions of the "average user" should not be the goal of design engineers.

5.9   Describe the effects of age, gender, and size on anthropometrics.

5.10 Identify three anthropometric data sources and explain how they should be used, the population they are most applicable to, and an industry that will benefit from this data.

REFERENCES

Ashdown, S., Loker, S., Schoenfelder, K.A., and Lyman-Clarke, L. (2004). Using 3D scans for fit analysis. *J. Textiles Apparel Manage. Technol.*, 4(1).

Center for Universal Design at North Carolina State University. (2011). http://www.ncsu.edu/project/design-projects/udi/

Davis, J., Valentine, T., and Davis, R. (2010). Computer assisted photo-anthropometric analyses of full-face and profile facial images. *Forens. Sci. Int.*, 200(1), 165–176.

Defence Technical Information center (DTIC), http://www.dtic.mil/dticasd/anthro.html (accessed September, 2011).

DOD-HDBK-743. (1980). *Anthropometry of U.S. Military Personnel*, U.S. Army Natick R&D Laboratories, DOD.

Eastman Kodak Company. (1983). *Ergonomics Design for People at Work*, Vol. 1, Van Nostrand Reinhold, New York.

Ergonomics Technologies Website. (2011). http://www.armscordi.com/SubSites/ERGO/ERG02_08.asp

Ergonomic Technologies. (2010). Business activities. Retrieved November 8, 2010, from Ergonomic Technologies: http://www.armscorbusiness.com/SubSites/ERGO/ERG02_08.asp

Garrett, J.W. and Kennedy, K.W. (1971). A collation of anthropometry. Technical Report AMRL TR-68-1, Aerospace Medical Research Laboratories: Wright-Patterson Air Force Base, OH.

Gordon, C.C., Churchill, T., Clauser, C.C., Bradtmiller, B., McConville, J.T., Tebbets, I., and Walker, R. (1989). 1988 anthropometric survey of US army personnel: Summary statistics interim report. United States Army Natick Research, Development and Engineering Center: Natick, MA.

Herzberg, F. (1968). Man and the nature of work.

Hung, C.Y.P., Witana, C.P., and Goonetilleke, R.S. 2004. Anthropometric measurements from photographic images. Paper presented at *Proceedings of the 7th International Conference on Work with Computer Systems (WWCS)*, Kuala Lumpur, Malaysia.

Hung, P., Witana, C., and Goonetilleke, R. (2004). *Anthropometric Measurements from Photographic Images in Work with Computing Systems* (Eds. H.M. Khalid, M.G. Helander, and A.W. Yeo), Damai Sciences: Kuala Lumpur, Malaysia, pp. 104–109.

Kroemer, K.H.E, Kroemer, H.J., and Kroemer-Elbert, K.E. (1990). *Engineering Physiology: Bases of Human Factors/Ergonomics*, 2nd edn., Van Nostrand Reinhold: New York.

Kroemer, K.H.E. (1994). *Locating the Computer Screen: How High, How Far? Ergonomics in Design*, January Issue, 40 and October 1993 Issue, 7–8.

Kroemer, K.H.E., Kroemer, H.B., and Kroemer-Elbert, K.E. (1994). *Ergonomics: How to Design for Ease and Efficiency*, Prentice Hall: Englewood Cliffs, NJ.

Kroemer, K., Kroemer, H., and Kroemer-Elbert, K. (2003). *Ergonomics: How to Design for Ease and Efficiency*, 2nd ed. Prentice Hall: Upper Saddle River, NJ.

Kroemer, K.H.E. (2006). *The Occupational Ergonomics Handbook*, Taylor & Francis: Boca Raton, FL.

Komlos, J. (1992). Anthropometric history: What is it? *Magazine of History*, 6(4), 3–5.

Kumar, S. (Ed.) (2008). *Biomechanics in Ergonomics*, 2nd edn., CRC Press: New York.

Lohman, T.G., Roche, A.F., and Martorell, R. (1988). *Anthropometric Standardization Reference Manual*, Vol. 10, Human Kinetics Books: Champaign, IL.

Loker, S., Ashdown, S.P., Schoenfelder, K.A., and Lyman-Clarke, L. (2005). Size-specific statistical and visual analysis of body scan data to improve ready-to-wear apparel fit. *J. Textiles Apparel. Manage. Technol.*, 5(1), 1–15.

Mace, R., Hardie, G., and Place, J. (1996). *Accessible Environments: Toward Universal Design*, Center for Universal Design, North Carolina State University: Raleigh, NC.

MIL-HDBK-759A (MI). (1981, June 30). *Human Factors Engineering Design for Army Material*, U.S. Army Human Engineering Lab, DOD.

NASA RP 1024. (1978). *Anthropometric Source Book: Volume 1: Anthropometry for Designers Anthropology Staff/Webb Associates*, NASA, Washington, DC, pp. 7–78.

NASA-STD-3000, (2000). Man-Systems Integration Standards, *Anthropometry and Biomechanics*, Volume I, Section 3. http://msis.jsc.nasa.gov/sections/sectionol.htm

National Institute of Standards and Technology Website, http://www.nist.gov/index.html

Pheasant, S. (1996). *Bodyspace: Anthropometry, Ergonomics, and the Design of Work*, Taylor & Francis: London, U.K.

Ressler, S. (n.d.). AnthroKids—anthropometric data of children. Retrieved November 7, 2011, from National Institute of Standards and Technology: http://ovrt.nist.gov/projects/anthrokids/

Roebuck, J.A. (1993). *Anthropometric Methods: Designing to Fit the human Body*, Human Factors and Ergonomics Society: Santa Monica, CA.

Roebuck, J.A., Kroemer, K.H.E., and Thomson, W.G. (1975). *Engineering Anthropometry Methods*, Wiley-Interscience: New York.

SAE International. (2010). CAESAR: Civilian American and European Surface Anthropometry Resource Project. Retrieved November 4, 2010, from SAE International: http://store.sae.org/caesar/

Sutherland, C.G. (1937). The rontgenographic image in the diagnosis of lesions of bone. *Brit. J. Radiol.*, 10(112), 295.

Wang, J., Thornton, J.C., Kolesnik, S., and Pierson, R.N. (2000), Anthropometry in body composition: An overview. *Ann. N. Y. Acad. Sci.*, 904, 317–326. DOI: 10.1111/j.1749-6632.2000.tb06474.x.

Wang, Y., Tussing, L., Odoms-Young, A., Braunschweig, C., Flay, B., Hedeker, D., and Hellison, D. (2006). Obesity prevention in low socioeconomic status urban African–American adolescents: Study design and preliminary findings of the HEALTH-KIDS study. *Eur. J. Clin. Nutr.*, 60(1), 92–103.

ANTHROPOMETRIC DATA SOURCES, RESOURCES, AND DATABASES

Databases

(*Source: http://www.dtic.mil/dticasd/anthro.html*)

- *AnthroKids*—Anthropometric Data of Children (1975 and 1977) Provides results of anthropometric studies of children undertaken by the Consumer Product Safety Commission in 1975 and 1977.
- *Anthropometric Data Sets*

Anthropometric Data Sets are a collection of civilian and military surveys spanning over 50 years. These data were formerly archived by the Air Force Armstrong Laboratory Human Engineering Division (now the U.S. Air Force Research Laboratory Crew System Interface Division) Center for Anthropometric Research Data. The Human Systems Information Analysis Center (HSIAC) turned the raw data into the Anthropometric Data Sets product available free for use.

* *Anthropometric Guidelines*—Department of Defense Ergonomics Working Group
A table of measurements and guidelines developed and published by the Department of Defense Ergonomics Working Group.
* *Anthropometry Source Book*—NASA Publication NASA-RP-1024 (1978)

  * Volume I
  Volume 1 presents all of the basic areas of anthropometry and its applications to the design of clothing, equipment, and workspaces for manned space flight.
  * Volume II
  Volume 2 contains data resulting from surveys of 61 military and civilian populations of both sexes from the United States, Europe, and Asia. Some 295 measured variables are defined and illustrated.
  * Volume III
  Volume 3 is an annotated bibliography covering a broad spectrum of topics relevant to applied physical anthropology with emphasis on anthropometry and its applications in sizing and design.

* *Anthropometry Survey of U.S. Army Personnel: Bivariate Frequency Tables*
In this report, bivariate frequency tables based on data from the 1988 Anthropometric Survey of U.S. Army Personnel are presented to facilitate the use of these data by designers of clothing, equipment, and workspaces which Army personnel will wear or use.
* *Anthropometry of U.S. Military Personnel (Metric) [DoD-HDBK-743A]*
The purpose of this handbook is to present body size information on the military personnel of the United States in the form of anthropometric data. These data are suitable for human factors engineering applications in the design and development of military systems, equipment, and material and in the design and sizing of military clothing and personal equipment.
* *Biodynamics Database (BDB)*
The Biodynamics Data Bank (BDB) was established in 1984 by a team of researchers at the Air Force Research Laboratory (AFRL), then known as the Armstrong Aerospace Medical Research Laboratory. The primary objective was to provide a national repository of Biodynamics test data accessible to the entire research community. The contents of the BDB include data from approximately 7000 acceleration impact tests conducted at the

Biodynamics and Acceleration Branch of AFRL, as well as information on the associated test facilities, test programs, and test subjects.

- *CAESAR (Civilian American and European Surface Anthropometry Resource Project)*: SAE International
  Whether designing new clothing lines or cockpits, accurate body measurement data is critical to create better and more cost-effective products. The CAESAR Product Line (CAESAR) was designed to provide the most current measurements for body measurements from a variety of countries. This product line was developed as a result of a comprehensive research project that brought together representatives from numerous industries including apparel, aerospace, and automotive.
- *Engineering Data Compendium*: Human Perception and Performance. Human engineering reference for systems design.
- *National Health and Nutrition Examination Survey (NHAMES)*
  The National Health and Nutrition Examination Survey (NHANES) run by the Centers for Disease Control (CDC) is a program of studies designed to assess the health and nutritional status of adults and children in the United States. It includes questionnaires, datasets, and protocols used in the series of studies.
- *Department of Defense Technical Reports*
  Anthropometry Research: R&D and Applied. These are a comprehensive current and historical listing of all publicly available DoD technical reports, and includes free electronic versions, where available, and instructions on how to obtain hard copies of individual reports.

Documents

- *Anthropometric Guidelines*—Department of Defense Ergonomics Working Group
  Anthropometric Guidelines. Males, Females, Population Percentiles.
- Anthropometry for Computer Graphics Human Figures—University of Pennsylvania
  Anthropometry as it applies to computer graphics is examined in this report that documents the anthropometry work done in the Computer Graphics Research Laboratory at the University of Pennsylvania from 1986 to 1988.
- Anthropometry for Persons with Disabilities: Needs for the twenty-first century
  This document is the final report of Task 2 of the "Anthropometric Research Review" undertaken by Anthropology Research Project, Inc. (ARP) for the U.S. Architectural and Transportation Barriers Compliance Board (Access Board) and administered by the U.S. Department of Education under Contract No. QA96001001. Under Task 1 of contract No. QA96001001, an annotated bibliography concerned with the anthropometry of people with disabilities and its applications to the design of facilities, workspaces, and equipment was completed. It appears in this report as an appendix.
- Anthropometry and Workspace Design—Cornell University (2007)
  Summary of guide to analysis of anthropometric data.

- Body Measurements (Anthropometry)—National Health and Nutrition Examination Survey III (NHANES III) (1988)
  Introduction to anthropometry, purpose, methodology, protocols, and equipment as applied to NHANES III.
- Ergonomic Models of Anthropometry, Human Biomechanics and Operator-Equipment Interfaces: Proceedings of a Workshop—National Academies Press (1988)
  Proceedings of the Ergonomic M, and Education (CBASSE).
- Gray's Anatomy
  Available as a free Web-based interactive document. The Bartleby.com edition of Gray's Anatomy of the Human Body features 1,247 engravings, many in color, from the classic 1918 publication, as well as a subject index with 13,000 entries.

6 Design of Workplaces and Hand Tools

6.1 LEARNING GOALS

The objective of this chapter is to provide an introduction to basic ergonomic principles of workplace design and evaluation of hand tools. The general process for workplace design will be studied. The student is also provided with sample evaluation tools and a detailed list of potential evaluation tools is also included.

6.2 KEY TOPICS

- Workplace design analysis
- Principles of arrangement
- General process for workplace design
- General principles for workplace design
- Designing to fit the moving body
- Designing for the standing operator
- Deigning for the sitting operator
- Working in confined and awkward spaces
- Designing for hand and foot operation
- Workplace evaluation tools
- Rapid upper limb assessment (RULA) tools
- Risk factors and risk prediction
- General guidelines for workplace controls and displays

6.3 INTRODUCTION AND BACKGROUND

Three hundred years ago, Ramazzini explained the reasons why he believed people who perform work while standing get more easily fatigued than people who have to work while walking. He alleged that in order to maintain an upright position, the muscles must be kept tensed. Also, he affirmed that workers who sit still and look down at their work in a stooped posture often become round shouldered and suffer from numbness in their legs, lameness, and sciatica. Ramazzini strongly believed that all sedentary workers suffered from lumbago (low back pain) and, therefore, recommended that workers should not consistently stand or sit still, but instead engage in dynamic activities to constantly move the body.

The work environment has changed drastically since the Industrial Revolution which began in the United Kingdom and spread to North America, and the rest of the world over 200 years ago. The American Industrial Revolution began in the early

1800s and produced substantial changes in agriculture, manufacturing, mining, transportation, and technology. As a result, the requirement for workers increased as did the need for the design of work environments. The work environment included many physically intensive tasks, and new equipment was designed to support the growing need in all industries, which were flourishing. However, there was not much emphasis on designing the work environment, equipment, or tools for humans. Around 1878, this began to change with work done by Frank and Lillian Gilbreth. As discussed in previous chapters, the Gilbreths' were pioneers in time and motion studies and conducted years of research and analysis to help workers in the industry. The objective of these studies was simplification and industrial efficiency. Much of their efforts improved the workplace for the operators and can be considered the beginning of applied ergonomics research. More than a hundred years later, ergonomists are still adapting to changes in technology and industry to design workplaces to accommodate the human operator. The objective of ergonomics in workplace design is to consider all elements of the physical task, information processing requirements, process, equipment, seating, and tools in order to create an integrated environment that is designed to meet the needs of the worker.

Ergonomically designed tools are more ubiquitous and affect a wider range of users than commonly known. Take the pencil, for example. The Romans were using metal styluses to write in 1300 BC. The earliest styluses, originally made of thin metal, eventually evolved into lead, which could be used to leave a light but readable mark on papyrus, but users did not know it was toxic. In 1564, a large graphite deposit was discovered in England, and graphite replaced lead because it left a darker mark and was nontoxic. However, graphite was so brittle that it needed a holder, so it was prudent to wrap sticks of graphite in string. Finally, a hollowed-out wooden case replaced the string, and by 1662 the first mass-produced wooden pencils were being sold in Nuremburg, Germany. This simple history is an example of how ergonomic problem solving was used to overcome various obstacles and imperfections in order to create a useful tool still used today (Pencil History, 2010).

6.4 WORKPLACE DESIGN ANALYSIS

The Occupational Safety and Health Administration (OSHA) states the following about ergonomics in workplace design (OSHA, 1987):

> The ergonomic approach to workplace design must be recognized as the most effective and is the first choice for controlling sources of workplace stress.

The reason for this is because the "best fit" is achieved by engineering the problems out at the design stage. In other words, focusing on task design and tool design rather than worker selection or training. With the implementation of the ergonomic approach to job design, prevention of injury is achieved as a result of the worker experiencing improved work postures, reduced forces, fewer repetitions, and reduction in overall exposure to risk factors.

In the design of workplaces, several analysis techniques can be considered (see summary of select analysis tools at the conclusion of this chapter). These investigation techniques can be used to analyze workplace dimensions, to determine human workplace interface, or to identify workplace factors that may contribute to ergonomic problems such as injuries, illnesses, operator discomfort, and reduced productivity. Some of these techniques include surveys, interviews, observation analysis workplace evaluation and analysis of tools and equipment used for the job task. Additionally, a variety of technology-based tools are available to support the collection, analysis, and interpretation of workplace or operator-related characteristics that affect job design.

6.4.1 Principles of Arrangement

In the design of a workplace there are often trade-offs as all desired conditions cannot generally be included. The principles of arrangement (Sanders and McCormick, 1987) should be followed to ensure a functional and well-designed workplace. These principles are as follows:

- Importance principle—The most relevant activities should lead the development of the design requirements.
- Frequency of use principle—Design such that frequently required activities are easily accessible, non-strenuous, and convenient.
- Functional principle—Group components according to the function they serve in the task.
- Sequence of use principle—If sequential patterns or routines are a part of task design, design such that the operator performs smooth transitions throughout the sequence.

6.4.2 General Process for Workplace Design

To effectively design a workplace, ergonomists should have significant details of the environment, task requirements, and anticipated worker population. A five-step approach for workplace design is summarized in the following (Figure 6.1):

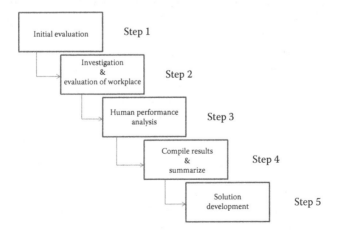

FIGURE 6.1 Five step approach to workplace design.

1. The first step in the workplace design analysis is to perform an initial evaluation.
   a. A thorough description of the environment and tasks needs to be meticulously studied, as well as any attributes that contribute to the task completion. Components of the analysis should include task and process design, workplace design, design of equipment, and tools.
   b. It is important to look for any job requirements that exceed the capabilities of the workers.
   c. Written procedures that workers are to follow should be evaluated for accessibility and ease of use.
   d. Procedures involving situational or external factors, policies, scheduling, and stress should also be evaluated.
2. The second step includes an investigation and evaluation of the workplace and environmental factors.
   a. Identify and evaluate all tools and equipment.
   b. Assess working conditions.
   c. Identify and evaluate safety.
3. The third step involves an analysis of human performance and physiological responses.
   a. Performance measures productivity over the shift. This can be measured in terms of total units completed, units produced per hour, idle time, time on arbitrary tasks, interruptions, distractions, and accidents.
   b. The quality of output is also measured, for instance the number of defects, the number of mistakes, and the number of times instructions must be referenced.
   c. Physiological measures that are studied during this step consist of direct measurements, such as heart rate, blood pressure, breathing rate, muscle activity (EMG reading), body or skin temperature, and oxygen consumption.
   d. Psychophysical measures, which are more subjective, are also considered. These include perceived exertion, difficulty, energy expended, and the level of fatigue.
4. The fourth step in the workplace design process is compilation of all results from the analysis.
   a. Summarize evaluation outcomes.
   b. Rank outcomes using a qualitative or quantitative scale (i.e., highly acceptable, acceptable, not acceptable).
   c. Prioritize factors of importance for inclusion in design.
5. The fifth step is solutions development, after sufficient information to develop strategies for design or redesign has been collected. The data should be used to develop ergonomic-based designs.
   a. It is useful to develop multiple design solutions that are at various levels of sophistication, costs, and time to implementation.
   b. Thus, it may be helpful to produce three levels of design solutions that include the following:

    i. Immediate design solutions
    ii. Interim design solutions
    iii. Long-term design solutions

Using the steps listed in the preceding text as a guideline will be useful in producing a broad workplace analysis; however, details of the work environment should be used as the guide to tailor this approach to the workplace needs. This process is designed to ensure that consideration is given to the person, workstation, work tasks, products, tools, and equipment. For instance, adjusting the workstation in order to reduce static muscle fatigue, minimize awkward postures, and improve visual acuity are considerations that address multiple aspects of the workstation. Adjusting work pieces or tools, such as jigs, clamps, and vises, is also an important consideration. These types adjustments help hold objects in place and minimize uncomfortable postures. Such adjustments can also reduce static loading, discomfort, application of force, and repetition of tasks. Efficient use of components and storage bins throughout the workplace eases accessibility of parts, reduces transportation time, distances, and storage space. Lift tables are highly recommended because they reduce lifting tasks, stooping, crouching, and twisting. A foot rest is another product that helps reduce fatigue and back strain, relieves stress on the legs, back, and neck, promotes circulation in the legs and feet, and promotes proper posture and comfortable seating. There are just a few examples of approaches that can be used to address worker needs.

6.5 GENERAL PRINCIPLES OF WORKPLACE DESIGN

Workplaces should be designed for safety, efficiency, and productivity. Ideally, this is done in the engineering design stage, prior to any operator exposure, as the engineering approach eliminates problems by designing them out of the environment. OSHA offers a resource in the ergonomic toolbox to provide guidance in varied areas of workplace design that include hand tool design and work station design (OSHA, 2010b). Excerpts from the NIOSH Ergonomic Toolbox are provided in Table 6.1.

In addition to applying ergonomic principles of workplace design, it is equally important to apply other principles to promote a safe, durable, and long-term workstation. A brief explanation and basis for ensuring the use of these principles is provided in the following:

- Maintainability
  - Increase product life.
  - Provide easy to understand instructions and procedures for replacing units or parts.
  - Consider accessibility or space available for the maintenance personnel to work.
  - Identify potential hazards (remove or warn users for hazards).
  - Provide warning labels and place them where they can be perceived, interpreted, and understood.
  - Follow established standards (i.e., ISO standards) for label placement and color.

TABLE 6.1
NIOSH Ergonomic Toolbox: General Workstation Design Principles

Tray 9–A. General Workstation Design Principles

1. Make the workstation adjustable, enabling both large and small persons to fit comfortably and reach materials easily.
2. Locate all materials and tools in front of the worker to reduce twisting motions. Provide sufficient work space for the whole body to turn.
3. Avoid static loads, fixed work postures, and job requirements in which operators must frequently hold static postures for long periods.
   a. Lean to the front or the side.
   b. Hold a limb in a bent or extended position.
   c. Tilt the head forward more than 15°.
   d. Support the body's weight with one leg.
4. Set the work surface above elbow height for tasks involving fine visual details and below elbow height for tasks requiring downward forces and heavy physical effort.
5. Provide adjustable, properly designed chairs with the following features:
   a. Adjustable seat height
   b. Adjustable up and down back rest, including a lumbar (lower back) support
   c. Padding that will not compress more than an inch under the weight of a seated individual
   d. Chair that is stable to floor at all times (five-leg base)
6. Allow the workers, at their discretion, to alternate between sitting and standing. Provide floor mats or padded surfaces for prolonged standing.
7. Support the limbs: provide elbow, wrist, arm, foot, and back rests as needed and feasible.
8. Use gravity to move materials.
9. Design the workstation so that arm movements are continuous and curved. Avoid straight-line, jerking arm motions.
10. Design so arm movements pivot about the elbow rather than around the shoulder to avoid stress on shoulder, neck, and upper back.
11. Design the primary work area so that arm movements or extensions of more than 15 in. are minimized.
12. Provide dials and displays that are simple, logical, and easy to read, reach, and operate.
13. Eliminate or minimize the effects of undesirable environmental conditions such as excessive noise, heat, humidity, cold, and poor illumination.

Tray 9–B. Design Principles for Repetitive Hand and Wrist Tasks

1. Reduce the number of repetitions per shift. Where possible, substitute full or semiautomated systems.
2. Maintain neutral (handshake) wrist positions:
   a. Design jobs and select tools to reduce extreme flexion or deviation of the wrist.
   b. Avoid inward and outward rotation of the forearm when the wrist is bent to minimize elbow disorders (i.e., tennis elbow).
3. Reduce the force or pressure on the wrists and hands:
   a. Wherever possible, reduce the weight and size of objects that must be handled repeatedly.
   b. Avoid tools that create pressure on the base of the palm that can obstruct blood flow and nerve function.
   c. Avoid repeated pounding with the base of the palm.
   d. Avoid repetitive, forceful pressing with the finger tips.

TABLE 6.1 (continued)
NIOSH Ergonomic Toolbox: General Workstation Design Principles

Tray 9–B. Design Principles for Repetitive Hand and Wrist Tasks

4. Design tasks so that a power rather than a finger pinch grip can be used to grasp materials. Note that a pinch grip is five times more stressful than a power grip.
5. Avoid reaching more than 15 in. in front of the body for materials:
   a. Avoid reaching above shoulder height, below waist level, or behind the body to minimize shoulder disorders.
   b. Avoid repetitive work that requires full arm extension (i.e., the elbow held straight and the arm extended).
6. Provide support devices where awkward body postures (elevated hands or elbows and extended arms) must be maintained. Use fixtures to relieve stressful hand/arm positions.
7. Select power tools and equipment with features designed to control or limit vibration transmissions to the hands, or alternatively design work methods to reduce time or need to hold vibrating tools.
8. Provide for protection of the hands if working in a cold environment. Furnish a selection of glove sizes and sensitize users to problems of forceful overgripping when worn.
9. Select and use properly designed hand tools (e.g., grip size of tool handles should accommodate majority of workers).

Tray 9–C. Hand Tool Use and Selection Principles

1. Maintain straight wrists. Avoid bending or rotating the wrists. Remember, bend the tool, not the wrist. A variety of bent-handle tools are commercially available.
2. Avoid static muscle loading. Reduce both the weight and size of the tool. Do not raise or extend elbows when working with heavy tools. Provide counter-balanced support devices for larger, heavier tools.
3. Avoid stress on soft tissues. Stress concentrations result from poorly designed tools that exert pressure on the palms or fingers. Examples include short-handled pliers and tools with finger grooves that do not fit the worker's hand.
4. Reduce grip force requirements. The greater the effort to maintain control of a hand tool, the higher the potential for injury. A compressible gripping surface rather than hard plastic may alleviate this problem.
5. Whenever possible, select tools that use a full-hand power grip rather than a precision finger grip.
6. Maintain optimal grip span. Optimum grip spans for pliers, scissors, or tongs, measured from the fingers to the base of the thumb, range from 6 to 9 cm. The recommended handle diameters for circular-handle tools such as screwdrivers are 3–5 cm when a power grip is required, and 0.75–1.5 cm when a precision finger grip is needed.
7. Avoid sharp edges and pinch points. Select tools that will not cut or pinch the hands even when gloves are not worn.
8. Avoid repetitive trigger-finger actions. Select tools with large switches that can be operated with all four fingers. Proximity switches are the most desirable triggering mechanism.
9. Isolate hands from heat, cold, and vibration. Heat and cold can cause loss of manual dexterity and increased grip strength requirements. Excessive vibration can cause reduced blood circulation in the hands causing a painful condition known as white-finger syndrome.
10. Wear gloves that fit. Gloves reduce both strength and dexterity. Tight-fitting gloves can put pressure on the hands, while loose-fitting gloves reduce grip strength and pose other safety hazards (e.g., snagging).

(continued)

TABLE 6.1 (continued)
NIOSH Ergonomic Toolbox: General Workstation Design Principles

Tray 9–D. Design Principles for Lifting and Lowering Tasks

1. Optimize material flow through the workplace by
   a. Reducing manual lifting of materials to a minimum
   b. Establishing adequate receiving, storage, and shipping facilities
   c. Maintaining adequate clearances in aisle and access areas
2. Eliminate the need to lift or lower manually by
   a. Increasing the weight to a point where it must be mechanically handled
   b. Palletizing handling of raw materials and products
   c. Using unit load concept (bulk handling in large bins or containers)
3. Reduce the weight of the object by
   a. Reducing the weight and capacity of the container
   b. Reducing the load in the container
   c. Limiting the quantity per container to suppliers
4. Reduce the hand distance from the body by
   a. Changing the shape of the object or container so that it can be held closer to the body
   b. Providing grips or handles for enabling the load to be held closer to the body
5. Convert load lifting, carrying, and lowering movements to a push or pull by providing
   a. Conveyors
   b. Ball caster tables
   c. Hand trucks
   d. Four-wheel carts

Tray 9–E. Design Principles for Pushing and Pulling Tasks

1. Eliminate the need to push or pull by using the following mechanical aids, when applicable:
   a. Conveyors (powered and non-powered)
   b. Powered trucks
   c. Lift tables
   d. Slides or chutes
2. Reduce the force required to push or pull by
   a. Reducing side and/or weight of load
   b. Using four-wheel trucks or dollies
   c. Using non-powered conveyors
   d. Requiring that wheels and casters on hand trucks or dollies have (i) periodic lubrication of bearings, (ii) adequate maintenance, and (iii) proper sizing (provide larger diameter wheels and casters)
   e. Maintaining the floors to eliminate holes and bumps
   f. Requiring surface treatment of floors to reduce friction
3. Reduce the distance of the push or pull by
   a. Moving receiving, storage, production, or shipping areas closer to work production areas
   b. Improving the production process to eliminate unnecessary materials handling steps
4. Optimize the technique of the push or pull by
   a. Providing variable-height handles so that both short and tall employees can maintain an elbow bend of 80°–100°
   b. Replacing a pull with a push whenever possible
   c. Using ramps with a slope of less than 10%

TABLE 6.1 (continued)
NIOSH Ergonomic Toolbox: General Workstation Design Principles

Tray 9–F. Design Principles for Carrying Tasks

1. Eliminate the need to carry by rearranging the workplace to eliminate unnecessary materials movement and using the following mechanical handling aids, when applicable:
   a. Conveyors (all kinds)
   b. Lift trucks and hand trucks
   c. Tables or slides between workstations
   d. Four-wheel carts or dollies
   e. Air or gravity press ejection systems
2. Reduce the weight that is carried by
   a. Reducing the weight of the object
   b. Reducing the weight of the container
   c. Reducing the load in the container
   d. Reducing the quantity per container to suppliers
3. Reduce the bulk of the materials that are carried by
   a. Reducing the size or shape of the object or container
   b. Providing handles or hand-grips that allow materials to be held close to the body
   c. Assigning the job to two or more persons.
4. Reduce the carrying distance by
   a. Moving receiving, storage, or shipping areas closer to production areas
   b. Using powered and non-powered conveyors
5. Convert carry to push or pull by
   a. Using non-powered conveyors
   b. Using hand trucks and push carts

Source: Adapted from design checklists developed by Ridyard, D., *Appl. Ergon. Technol.*, pp.19046–3129; DHHS (NIOSH) publication number 97–117, http://www.cdc.gov/hiosh/docs/97–117/pdfs/97–117.pdf website, accessed April 18, 2011.

- Safety
  - Safety should be a priority design.
  - Provide guards, fail-safe mechanisms, warnings, labels, and alarms.
  - Safety should not impair productivity.
  - Provide protection against accidental activation of control switches.
- Usability
  - Apply usability principles in design.
  - Increase efficiency and minimize errors.
  - Consider human capabilities, human stature dimensions, and population anthropometrics.
  - Provide clear, concise documentations and instructions for users.
- Comfort
  - Reduce fatigue, complaints, and discomfort.
  - Consider expected use and possible misuse by the worker.
  - Promote user acceptance.
  - Prototype testing.

- Reliability
  - Increase use and confidence in workstation.
  - Support quality and scheduling.
- Durability
  - Design to withstand expected factors related to users' expected lifespan.
  - Design with consideration of application environment.

6.5.1 DESIGNING TO FIT THE MOVING BODY

Humans are dynamic and function better over extended periods when allowed to vary posture, loading, and work position. The workplace should be designed to accommodate a range of motions that are expected of the operator for task performance. One way to accomplish this is to select the extremes of body positions that are expected and design for the motions between these ranges.

Traditionally, there are three work postures or combinations of these postures including lying down, sitting, and standing. However, there are other positions, for example, kneeling on one or both knees, squatting, and stooping. In order to measure the effects of work postures, both independent and dependent variables need to be considered. Independent variables relate to body movements, postures, and change measurement variables, such as rate of change, range of change, and maximum displacement. Dependent variables fall within a range of disciplines including physiological, for example, oxygen consumption, heart rate, blood pressure, and electromyogram; medical, such as musculoskeletal disorders (MSDs) and cumulative trauma disorders (CTDs); biomechanical, for instance spine disc pressure, x-rays, CAT scans, and force calculations from models; engineering, for example, productivity, forces, and pressures on seat, backrest or floor; and psychophysical, including subjective measures, such as interviews and surveys (Kroemer et al., 1994). Subjective measures consist primarily of subjective pain or discomfort ratings. Subjective measures have been highly correlated with many physiological measures, such as heart rate and psychological rating of perceived exertion. However, these measures are highly dependent on how well the survey is administered and the results can be quite accurate.

Biomechanics is a useful tool to evaluate dynamic task performance, as the field studies the human body in movement. These types of evaluations are used extensively in occupational analysis and to enhance the performance of athletes, particularly in elite athletics such as the Olympics. Biomechanical analysis can include the use of modeling tools video analysis, and mathematical models to predict the loading and impact of a work environment on an operator, indirect and direct measurement tools to assess capabilities in specific postures, or task activities throughout task performance.

6.5.2 DESIGNING FOR THE STANDING OPERATOR

Standing is usually used as a working posture if sitting is not possible, because the operator must exert large forces with the hands, the work area requires mobility, or the operator has to cover a large work area. Standing in one place should only

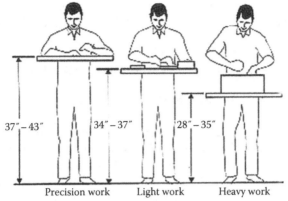

Precision work   Light work   Heavy work
Different tasks require different work surface Heights—NIOSH

FIGURE 6.2   Standing workstation design dimensions.

be imposed for a limited time period. Forcing someone to stand because the work area is placed high above the floor or outside the sitting reach areas is not a good justification.

The working height is critically important in the design of workplaces, and this height depends on the activity and the size of the object handled such as hand tools (Figure 6.2). The main reference point is the elbow height of the worker, because generally the strongest hand forces and the most useful mobility are between the elbow and hip heights. If the working height is too high, the shoulders must be frequently lifted, causing discomfort and painful cramps in both the neck and shoulders. Working heights at or above shoulder height can also contribute to shoulder tendonitis or other cumulative trauma disorder (CTDs) of the shoulder. Likewise, if work is placed too low, the back must be constantly bent leading to temporary or potentially long-term discomfort, pain, or injury.

The average elbow height (distance from floor to underside of elbow when it is bent at right angles with the upper arm vertical) is almost 1070 mm for men and 1000 mm for women both in North America and Europe. The generally recommended working height for hand work, while standing, is between 50 and 100 mm below the elbow level. The following text lists three recommended height guidelines when performing certain types of standing work (Kroemer and Grandjean, 1997):

- Delicate work or fine manipulation (i.e., drawing)
  - Support the elbows to minimize loading on the back.
  - Design to support viewing requirements.
  - Design for work surfaces 50–100 mm above the elbow height.
- Manual work (i.e., light assembly)
  - Allow space for equipment, hand tools, materials, and containers.
  - Design for work surfaces 100–150 mm below the elbow height.
- Exertion of extensive manual effort or constant use of weight (i.e., heavy assembly or woodworking)

- When constant use of the strength of the upper part of the body is involved (heavy assembly work or woodworking), the working surface needs to be lower.
- Provide forearm and wrist support when possible.
- Design for work surfaces 150–400 mm below the elbow height.

Table 6.1 provides more detailed dimensions for designing standing workstations for men and women.

Sufficient room for the feet of the operator must also be provided, including toe and knee space in a standing workplace. This space allows the worker to move closer to the workspace. The floor at the workstation should be flat and free of obstacles to reduce the risk of tripping. Likewise, sufficient friction between the soles of shoes and floor surfaces is necessary to reduce the risk of slips and falls. Finally, to promote comfort when standing, elastic floor mats and soft soles have been useful for reducing foot, leg, and back discomfort.

To manage the static loading that results from standing workstations, employers will sometimes provide stand seats or semi-seats. These semi-seats should provide a level of support between standing and sitting without interfering with the task performance. Due to variability within a task, this type of seating is often useful for certain job activities without impacting overall task performance. However, it is important to remember to not fully take the weight off the legs using these seats, as they tend to be higher than traditional seating and oftentimes workers cannot fully reach the floor and thus can lead to ischemia (occlusion of blood flow in the legs). Possible disadvantages that come with these seats include stability issues, restriction of the range of motion, and discomfort due to ischemia in the lower limbs.

6.5.3 Designing for the Sitting Operator

Sitting down is a much less strenuous working posture than standing. Stabilizing the body while sitting down is easier because fewer muscles need to be contracted and the body benefits from the support of the chair. Sitting allows better control of hand movements, but the seated postures mean the capacity for force exertion with the hands is reduced. Sitting individuals can operate controls with their feet; however, the amount of force that can be applied is considerably less than when standing. When designing a workstation for a sitting operator, it is important to consider the free space required by the legs and feet for movement in all directions. Also, the seat should be designed so that the sitting posture can be changed frequently. An adjustable chair that meets ergonomic design criteria should be considered. An overview of the dimensions of interest for seated workers is shown in Figure 6.3.

Humans are not biomechanically adapted for long-term static postures, whether it is standing or sitting. Therefore, sitting for extended periods of time can lead to lower back pain and foot swelling. Changing postures throughout the shift will reduce continued compression of the spinal column, as well as muscle fatigue. The preferred working area is right in front of the body at about elbow height, with the upper arms hanging. Visual requirements should also be considered when determining work

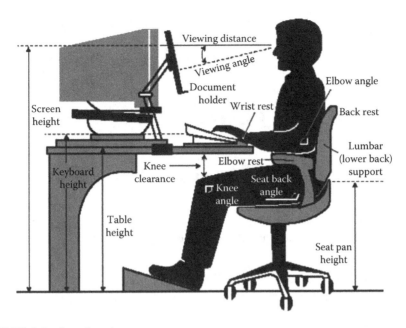

FIGURE 6.3 Seated worker.

area height. Seat heights should range from 35 to 50 cm, and the comfort of the seat is determined by the design of the seat pan, backrest, material that touches the body, and adjustability.

6.5.3.1   Elements of an Ergonomic Chair

When picking out a chair, it is critical to consider the type of chair, lumbar support, foot rests, and appropriate height level. Many physical and task requirement factors can produce risk factors for back pain, as evidenced by the fact that the primary complaint of sedentary workers is back problems. According to OSHA, the chair is an essential element of workstation design (OSHA, 2011). Ergonomic seating requirements include the following:

- Lumbar support
  - Supports the lower region of the back
  - Promotes proper posture
  - Minimizes the load on the lower back
- Chair with a waterfall edge
  - Results in termination of seat in back of thigh or lower leg
  - Prevents the obstruction of blood flow to the lower extremities
- Armrest
  - Adjustable, compatible with task performance
- Slightly tilted seat pan
  - Promotes proper posture, particularly in the lower back

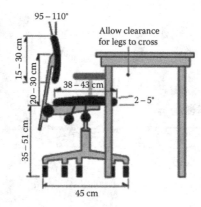

FIGURE 6.4 Ergonomic chair range of dimensions. (Source: Canadian Centre for Occupational Health and Safety, http://www.ccohs.ca)

- Adjustability
  - Seat pan adjustment
  - Back rest
  - Height

The qualities of a safe and healthy chair are shown in the OSHA eTools as well as on the Canadian Centre for Occupational Health and Safety (CCOHS) website (Figure 6.4).

The tools contained in the OSHA eTools were developed by the U.S. Federal Government under the direction of a technical editorial board lead by Brett Besser (OSHA eTools, 2011). It is important that we move our bodies frequently at work and avoid static postures for long periods of time. The traditional thinking that everyone should sit completely upright is mistaken. Many different postures can be comfortable, and there is no one "best" posture. Work furniture should be designed to support task requirements and placed for easy variations of postures. Sitting on flat hard surfaces causes the natural lordosis of the spine to become flattened. Therefore, sitting on such surfaces should be avoided. Also, forward declining seats force the hip angle to open and contribute to lordosis. Some of the disadvantages that come with forward declining seats include: hard to sit on, forward slippage, shin pressure, increased fatigue, and uneasy work positions. Comprehensive guidelines for office ergonomics is contained in the OHSA eTools resource (OSHA, 2011).

6.5.4 WORKING IN CONFINED AND AWKWARD SPACES

Confined space operations exist throughout a vast majority of industries and in some cases have resulted in injuries and deaths. Examples of confined work spaces are shown in Figure 6.5. According to the Bureau of Labor Statistics (BLS), from July 1, 1972, to January 2002, 119 workers were injured and 353 were killed in confined spaces (BLS, 2004). John Rekus, a well-known researcher in the area of confined space operations, attributes confined space injuries and deaths to atmospheric or

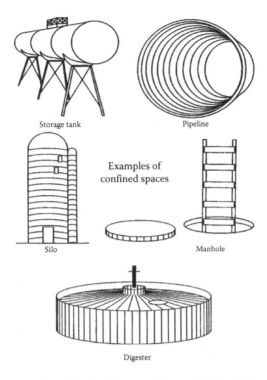

Storage tank

Pipeline

Examples of
confined spaces

Silo

Manhole

Digester

FIGURE 6.5 Examples of confined work spaces (Source: Pettit and Linn, A Guide to Safety in Confined Spaces, 1987, CDC website: http://www.cdc.gov/niosh/topics/confinedspace/).

oxygen-deficient environments, explosions, becoming trapped, electrocution, becoming caught or crushed, being struck by falling objects, ingress or egress issues, temperature extremes, noise vibrations, and stress from exertion (Rekus, 1994). These risk factors are often a result of task design constraints (i.e., underground mining) or inherent task risks, but in some cases result from a lack of appropriate ergonomic and safety practices. While some issues in task requirements cannot be changed, the use of ergonomics and safety to support workers in confined spaces should be a priority in reducing operator risk levels. Confined space workers are also subject to psychological risk factors, such as claustrophobia, maintaining appropriate attention levels, and situational and spatial awareness. These psychological operator needs should be addressed as an aspect of task design.

On January 14, 1993, the U.S. OSHA adopted the Final Rule for Permit-Required Confined Spaces for General Industry (29 CFR 1910.146) with an effective implementation date of April 15, 1993. This regulation requires employers to set up a comprehensive confined space program that includes, but is not limited to, identification, permitting, testing, training, emergency response, and rescue (OSHA Directives, May 5, 1995) (Figure 6.6).

Although there has been some research in the areas of work design, workload, task analysis and habitability pertaining to confined environments, including the Arctic underwater mining, NASA studies, and tanking operations, there is a need

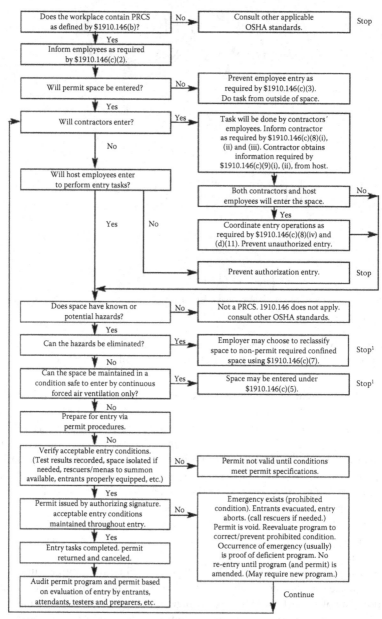

¹Spaces may have to be evacuated and reevaluated if hazards arise during entry.
Source: 29 CFR 1910.146 appendix A.

FIGURE 6.6 Permit required confined space decision flow chart.

for further research in order to identify the risk factors that affect human performance in all types of confined space industries—not only in developing a formal risk factor matrix, but ultimately in utilizing the identified risk factors in mitigating potential injury to the confined space worker (Brody, 1993; Randolph, 1996; Relvini and McCauley-Bell, 2005; Winn et al., 1996).

6.5.5 Designing for Hand Use

The human hand is one of the most intricate members of the human anatomy containing over 29 major and minor bones, 29 major joints, 48 nerves, and 123 ligaments. This network of joints and ligaments enable the hand to assume a broad array of postures and movements. In the occupational environment, the hand is used extensively in physical task performance and the human hand is capable of performing a large variety of required activities. These activities range from fine manipulation to coarse, forceful movements and task performance can be categorized according to Kroemer et al. as follows (Kroemer et al, 2010):

- Fine manipulation of objects, requiring little displacement and force, for example, writing, adjusting controls, and assembling small parts
- Fast movements toward an object, requiring moderate accuracy to reach the target but fairly small exertions of force, for example, moving toward a switch and then operating it
- Frequent movements between targets that require higher accuracy but little force, for example, an assembly task that involves taking parts from bins and assembling them
- Forceful activities with little or moderate displacement, for example, turning a hand tool against resistance
- Forceful activities with large displacement, for example, using a hammer to put a nail in the wall

The three categories of requirements when designing tasks for hand use are accuracy, strength, and displacement. *Fitts' Law* (Fitts, 1954) has been used historically as a foundational source for guidance in design for accurate and fast movements. Fitts' Law was developed to predict hand movement time and can be used to promote accuracy in task performance. The premise of this law is that movement time is a direction function of task difficulty, and that task difficulty is directly related to movement distance and inversely related to the size of the target. The following equation describes this relationship:

$$MT = a + b \log_2 \left(\frac{2A}{W} \right) \qquad (6.1)$$

where
  MT is the movement time
  a and b are empirically determined constants
  A is the movement amplitude (distance) from start to centerline of target
  W is the width of the target

The term $\log_2(2A/W)$ is referred to as the index of difficulty and is a measure of the difficulty of the motor task.

This law should be considered in task design involving hand tasks and hand movements particularly for precise activities.

Fitts' Law has also been applied to foot motion (Drury, 1975). This study used a modification of this law to predict movement time for operating foot pedals. Drury defined the index of difficulty as follows:

$$ID = \log_2\left(\frac{A}{W} + S\right) + 0.05 \qquad (6.2)$$

where
    ID is the index of difficulty
    A is the movement amplitude (distance) from start to centerline of target
    W is the width of the target
    S is the shoe sole width

This information can be used in design foot controls, pedals, or other foot-activated devices.

Methods-time measurement (MTM) is a well-researched area of occupational ergonomics, laboratory experiments, and process design. This work was originally referred to as scientific management (Luczak, 1997). MTM is a procedure for improving methods and establishing time standards by recognizing, classifying, and describing the motions used or required to perform a given operation and assigning predetermined time standards to these motions (MTM.org, website, 2011). The MTM Association for Standards and Research is a nonprofit organization that has led the way in the development of computerized and manual work measurement systems that recognize, classify, describe, and objectively measure the performance of individuals working at various levels within an organization (MTM.org, website, 2011). Although this area began with one method for time measurement, it has evolved into a dynamic family of systems that can be used in numerous industrial settings. An example of the use of MTM times is shown in Table 6.2 (Maynard, 1948). This table can be used to predict movement time for specific distances given a particular type of movement (Luczak, 1997).

6.5.5.1 Hand Tool Design

The design of hand tools is a complicated ergonomic task because several factors must be simultaneously considered. It is essential that hand tools fit the contours of the hand while using strength and energy capabilities required for task performance without overloading the body; likewise tools should be designed so that they can be held securely with a straight wrist and proper arm posture. Numerous disorders are associated with improper use of hand tools when ergonomic principles are violated. Some disorders that have been reported due to non-powered and powered hand tools include (NIOSH, 1995) the following:

TABLE 6.2

Example of a MTM Table of Motion Times

Methods-Time Data

Table I-Reach

			Leveled Time TMUs					
Case	Description	Distance moved (in.)	A STD.	A Hand in Mot.	A with C, D or B	B Hand in Mot.	C or D	E
A	Reach to object in	1			2.1		3.6	
	fixed location or to	2			4.3		5.9	
	object in other hand	3			5.9		7.3	
	or on which other hand rests	4	6.1	4.9	7.1	4.3	8.4	6.8
B	Reach to single object	5	6.5	5.3	7.8	5.0	9.4	7.4
	in location that may	6	7.0	5.7	8.6	5.7	10.1	8.0
	vary slightly from	7	7.4	6.1	9.3	6.5	10.8	8.7
	cycle to cycle	8	7.9	6.5	10.1	7.2	11.5	9.3
C	Reach to object in	9	8.3	6.9	10.8	7.9	12.2	9.9
	group	10	8.7	7.3	11.5	8.6	12.9	10.5
		12	9.6	8.1	12.9	10.1	14.2	11.8
		14	10.5	8.9	14.4	11.5	15.6	13.0
D	Reach to very small	16	11.4	9.7	15.8	12.9	17.0	14.2
	object or where	18	12.3	10.5	17.2	14.4	18.4	15.5
	accurate grasp is	20	13.1	11.3	18.6		19.8	16.7
	required	22	14.0	12.1	20.1		21.2	18.0
E	Reach to indefinite	24	14.9	12.9	21.5		22.5	19.2
	location to get hand	26	15.8	13.7	22.9		23.9	20.4
	in position for body	28	16.7	14.5	24.4		26.3	21.7
	balance or next motion or out of way	30	17.5	15.3	25.8		26.7	22.9

Source: Maynard et al. (1948).

- Acute musculoskeletal problems
- Muscle, tendon, or ligament tear, bone fractures
- Chronic Musculoskeletal Disorders (MSDs)
- Vascular disorders
- Vibration "white finger"
- Hearing impairments
- Respiratory disorders

The types of industries in the United States that are prone to occupational injuries range from service industries to agriculture. Some of the primary industry divisions experiencing nonfatal occupational injuries due to hand tools in the United States are listed in Table 6.3 (U.S. Bureau of Labor Statistics, 2002).

TABLE 6.3

Nonfatal Occupational Injuries Due to Hand Tools in Major U.S. Industry Divisions (2002)

Industry	Total Number	Due to Hand Tools
All private industry	1,436,194	66,588
Agriculture, forestry, and fishing	31,520	2,185
Mining	11,355	758
Construction	163,641	14,439
Manufacturing	280,005	16,032
Transportation and public utilities	168,632	3,116
Wholesale and retail trade	372,192	19,179
Finance, insurance, and real estate	36,689	1,399
Services	372,159	9,443

Source: Kumar, S. (Ed.), *Biomechanics in Ergonomics*, 2nd edn., CRC Press, New York, 2008.

While the actual number of incidents is larger in some industries such as services, the evidence that this is more of an issue in physically intensive industries such as construction, transportation, mining, and agriculture is seen when reviewing the incidence rates. Table 6.4 provides incidence rates for nonfatal occupational industries due to hand tools in the United States.

Given these statistics, an emphasis should be placed on designing safe hand tools, procedures for use, and employing necessary administrative controls to reduce risk of injury in hand tool use.

Hand grip on a tool, also known as tool coupling, is important in defining the amount of force transferred, as well as the precision of transfer. Kroemer identifies ten generally recognized tool couplings as follows (Kroemer, 1986, 2001):

1. Digit touch
   a. One digit (finger) touches an object
2. Palm to touch
   a. Some part of the palm or hand touches object
3. Finger palmar grip (hook grip)
   a. One finger or several fingers hook onto a ridge or handle. This type of finger action is used where thumb counterforce is not needed.
4. Thumb-fingertip grip (tip pinch)
   a. The thumb tip opposes one fingertip.
5. Thumb-finger palmar grip (pad pinch or plier grip)
   a. Thumb pad opposes the palmer pad of one finger (or the pads of several fingers) near the tips.
6. Thumb-forefinger side grip (lateral grip or side pinch)
   a. Thumb opposes the radial side of the forefinger

TABLE 6.4

Incidence of Nonfatal Occupational Injuries in the United States (2002) Due to Different Types of Non-Powered Hand Tools Classified by Major U.S. Industry Divisions

Tool Type	Total Number	Agriculture, Forestry, and Fishing	Mining	Construction	Manufacturing	Transportation and Public Utilities	Wholesale Trade	Retail Trade	Finance, Insurance, and Real Estate	Services
A	162.6	208.6	198.8	276.8	174.5	270.6	175.0	153.6	53.5	133.8
B	10.6	16.2	15.6	30.6	12.0	12.3	8.0	10.8	3.0	7.0
C	5.1	9.4	11.9	13.5	5.8	3.6	3.9	7.9	1.6	2.3
D	0.1	—	0.2	0.3	0.1	—	0.1	—	—	0.0
E	2.7	5.2	0.5	5.1	2.3	0.7	2.1	6.3	0.8	1.2
F	0.5	2.0	—	3.0	0.4	0.7	0.2	0.1	0.4	0.3
G	0.1	—	3.0	0.2	0.3	0.2	—	—	—	0.1
H	0.1	—	—	—	0.1	0.0	—	0.1	—	0.1
I	0.5	0.6	2.3	2.6	0.9	0.5	0.5	0.1	0.2	0.1
J	0.1	—	—	0.2	0.1	—	—	—	—	0.0
K	0.4	0.6	2.8	0.9	0.9	0.7	0.6	0.2	0.1	0.1
L	0.6	0.8	2.9	1.1	0.6	0.7	0.3	1.0	0.1	0.4

Source: Kumar, S. (Ed.), Biomechanics in Ergonomics, 2nd edn., CRC Press, New York, 2008.

Note: A, total in all private industries; B, tools, instruments, and equipment; C, non-powered hand tools; D, non-powered boring hand tools; E, non-powered cutting hand tools; F, non-powered digging hand tools; G, non-powered gripping hand tools; H, non-powered measuring hand tools; I, non-powered striking and nailing hand tools; J, non-powered surfacing hand tools; K, non-powered turning hand tools; L, non-powered hand tools such as crowbars, pitchforks, etc.

7. Thumb-two-finger grip (writing grip)
   a. Thumb and two fingers, often forefinger and middle finger, oppose each other at or near the tip.
8. Thumb-fingering enclosure (disc grip)
   a. Thumb pad and the pads of three or four fingers oppose each other near the tips (object grasped does not touch the palm). This grip evolves from coupling #7.
9. Finger-palm enclosure (collet enclosure)
   a. Most, or all, of the inner surface of the hand is in contact with the object while enclosing it. This enclosure evolves easily from coupling #8.
10. Power grasp
   a. The total inner hand surface is grasping the (often cylindrical) handle that runs parallel to the knuckles and generally protrudes on one or both sides from the hand. This grasp evolves easily from coupling #9.

Although there are often additional risks when using powered hand tools, the use of non-powered hand tools poses definite risks in task performance. These tools can include hammers, crowbars, pitch forks, and digging tools. The incidence rate for nonfatal occupational injuries in the United States due to non-powered hand tools is shown in Table 6.4.

Incidence rates represent the number of injuries and illness per 10,000 full-time workers and are computed as follows:

$$\text{Incidence rate} = \left(\frac{N}{EH} \times 20,000,000 \right) \qquad \text{(Figure 6.3)}$$

where
   N is the number of injuries
   EH is the total hours worked by all employees during the calendar year

20,000,000 is the base for 10,000 equivalent full-time workers (working for 40 h/week, 50 weeks/year).

Hand tools are either manual, power driven, or a combination of both. When using manually driven tools, the operator generates all the energy and is always in full control of the energy exerted, except with percussion tools. A few examples of manually driven tools and tasks are described in the following:

- *Percussive*: swing or hold handle
- *Scraping*: push, pull, or hold handle
- *Rotation or boring*: push, pull, turn, or hold handle
- *Squeezing*: press and hold handle
- *Cutting*: pull or push while holding handle, for example, scissors and knives

When handling power-driven tools, the operator holds places and moves the tool. Energy is generated by an external power source with the exception of the operators' dynamic static load in holding the tool. Sudden resistance can cause injury and vibrations can lead to overuse disorders or CTDs. Proper posture of the hand is important when handling hand tools. The wrist should never be bent but should be kept straight to avoid overexertion of the muscles and hand tissues, as well as compression of nerves and blood vessels. Normally, the grasp centerline is at about 70° to the forearm axis. It is always better to bend the tool than to bend the wrist. Handles in tools need to be designed so that the handle fits to the form of the hand, instead of having a straight and uniform surface. Bulges and restrictions generate a form fit but can make the grip uncomfortable, however, when used in a manner consistent with hand anatomy. Friction on surfaces can also help improve grip and reduce the risk of dropping the tools. The following rules should be applied when designing or working with hand tools (Kroemer, 2001):

- Push or pull in the direction of the forearm, with the handle directly in front of it.
- Keep the wrist straight.
- Provide good coupling between hand and handle, by shape and friction.
- Avoid pressure spots and pinch points.
- Avoid tools that transmit vibrations to the hand.
- Do not operate tools frequently and forcefully by hand; automated interaction is preferable.

When considering existing tools in a workplace, the ergonomic impact of the design on the operators' hand should be considered. In study by Dababneh et al. (2004), on evaluation of non-powered hand tools, 18 tools were selected to represent typical non-powered hand tools commonly used in the construction and carpentry trades. A 16-item checklist was developed that can be useful in hand tool evaluation (see Figure 6.7).

6.5.5.2 Power Hand Tools

In 1994, the Engineering Control Technology Branch (ECTB) of the National Institute for Occupational Safety and Health initiated a project entitled the "Ergonomic Study of Power Hand Tools in the Automotive Manufacturing Industry" due to a prevalence of work-related MSDs. According to this study, the ergonomic evaluation of power tools categorizes and evaluates properties that include

- Tool weight and load distribution
- Tool grip size, shape, and texture
- Torque reaction
- Torque scatter performance
- Vibration

Vibration is a factor that has had substantial impacts on workers. This includes low-, moderate-, and high-frequency vibration. Some hand-held power tools can produce substantial vibration and in some cases tools generate vibration levels so high they can damage the blood vessels and nerves in the operators' hands. Workers are at a greater risk of developing a vibration-related injury if they use vibrating power tools often or for long periods of time. The hand-arm vibration syndrome (HAVS) is the

Checklist for the Ergonomic Evaluation of Hand Tools

Considering the job the tool is designed for and the work environment, respond to each item on the checklist by "Yes," "No," or "NA" (not applicable). Place the score that corresponds to your response in the "score" column. Add the scores of all items to get the total score of the tool. Maximum score is 100.
(*Items 7, 8, 9, and 10 are not applicable for all tools.)

First thing first: The tool will do the job with the desired quality and will last as expected:
Yes: continue with the checklist;
No: reject the tool.

Item	Ergonomic Feature	Yes	NA	No	Score
1	Grip surface is non-slippery.	+10		0	
2	Grip surface does not have sharp edges, undercuts, deep ribs, and/or finger grooves.	+10		0	
3	Grip surface is electrically insulated; tool handle is either made of wood or coated with rubber or soft plastic.	+10		0	
4	Grip surface is thermally insulated; it will not get hot or cold quickly when working in a hot or cold environment.	+2		0	
5	Handle is made of wood, or grip surface is coated with semi-pliable material; not too hard and not too soft, similar to the rubber used in the soles of sport shoes.	+10		0	
6	Grip length is 4 – 6"; handle does not end inside the palm of the hand.	+10		0	
7	For one-handle tools: Size of handle cross section is not too small or too large. The index finger and the thumb are allowed to overlap by 3/8" when gripping (for hammers and hammer-like tools, overlap of 1" is acceptable).	+8*	0	0	
8	For one-handle tools other than screwdrivers: Shape of handle cross section is oval or rounded-edge rectangular.	+2*	0	0	
9	For screwdrivers: The basic shape of handle cross section is circular, hexagonal, square, or triangular.	+2*	0	0	
10	For two-handle tools (plier-like): Grip span is greater or equal than 2" when fully closed and less than or equal to 3.5" when fully open.	+10*	0	0	
11	Angle of the handle is formed so that the work can be done keeping a straight wrist.	+10		0	
12	The tool weight is less than 5 lb.	+10		0	
13	The tool can be used with either hand.	+2		0	
14	The tool can be used with the worker's dominant hand.	+10		0	
15	The tool will allow a two-handed operation (using both hands at the same time).	+4		0	
16	The tool and accessories are clearly marked and/or color coded so they are easy to identify; colors are bright and tool contrasts with the surroundings of the work area.	+2		0	
	Total score of the tool (100 points possible)				

*Measurement scale (inches)

FIGURE 6.7 Checklist non-powered hand tools.

result of regular use of vibrating pneumatic, electric, hydraulic, or gasoline-powered hand tools. This musculoskeletal condition can be permanently disabling.

Power hand tools generally use electric motors but can also be pneumatic, hydraulic, gas, battery, or power actuated. A few examples of power hand tools include

- Drills
- Staple guns
- Pneumatic torque wrenches
- Nail guns
- Portable abrasive wheel tools
- Floor sanders

Operators using power hand tools are often exposed to the hazards of falling, flying, abrasive and splashing objects; or exposed to harmful dusts, fumes, mists, vapors, or gases. The appropriate personal protective equipment is necessary to protect from the hazard. Hazards and ergonomic risk factors involved in the use of power tools can be prevented by the following five basic safety rules (Hand and Power Tools, 2011; OSHA, 2011):

- Keep all tools in good condition with regular maintenance.
- Use the right tool for the job.
- Examine each tool for damage before use.
- Operate according to the manufacturer's instructions.
- Provide and use the proper protective equipment.

6.5.6 DESIGNING FOR FOOT OPERATION

Compared to hand movements over the same distance, foot motions consume more energy, are less accurate, and are slower. However, they are considerably more powerful. If the person stands at work, little force and infrequent use of the foot should be used, because these physical efforts require the operator to stand on the other leg alone. For a person who sits while at work, the operation of foot controls is not as complicated, because the seat supports the body, allowing the feet to move as needed.

The amount of force that can be exerted by the foot depends largely on the body posture during actuation. Following are some rules that are used in the design of foot controls (Helander, 2006):

- Require repeated foot operation only from a seated operator.
- Design for pushing roughly in the direction of the lower leg.
- Small forces should be applied by tilting the foot about the ankle.
- Perform large forces by pushing with the whole leg, if possible with a solid back support from the seat.
- Do not require fine control, continuous operations, and quick movements simultaneously.

For the sitting operator of a foot control, guidelines suggest that light downward forces are best produced at knee angles of 105°–110°, while strong forward forces require knee angles of 135°–155°. For the standing operator, the controls and work station should allow the knee to bend comfortably; tasks should not overstress the foot through excessive repetition in actuation, and should be within reach of the foot work envelope.

6.6   WORK ENVELOPES

The primary work space (i.e., work surface) should take into account the maximum vertical and horizontal reach of both male and female workers. This space, defined as the work envelope, represents the space where task activities, equipments, and tools should be placed for an individual to perform the task without excessive reach, awkward posture, or discomfort. An example of seated work reach envelopes are provided in Figure 6.8. In the standing work envelope, the individual should not have to bend forward and only the extension of the arm and hand should be considered as the functional work space is designed. The work stations should be designed for the fifth percentile female to reach the work area and the most frequently performed activities should be performed while the individual's elbows are at the sides, thus creating a smaller vertical and horizontal work envelope (Eastman Kodak, 1983). The dimensions used to define the work envelopes should be compatible with the employee population and take into consideration all work, equipment, and tools that

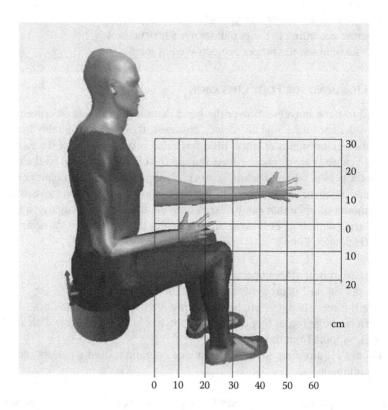

FIGURE 6.8   Example of a worker reaching for envelopes.

are required for task performance. For a detailed listing of dimensions, see Chapter VI, Appendix A of Eastman Kodak (1983).

6.7 WORKPLACE EVALUATION TOOLS

A useful resource to identify applicable ergonomic tools is *The Occupational Ergonomics Handbook* (Marras and Karwowski, 2001). Likewise, Dr. Thomas Bernard in the College of Public Health at the University of South Florida has compiled an impressive summary of ergonomic analysis tools (Bernard, 2011). A number of postural and workplace evaluation tools are available to support an ergonomic analysis. An abbreviated list of some of these tools is provided at the conclusion of this chapter. Any tool that is used should follow sound ergonomic principles, utilize accepted data or norms, and have undergone some form of independent evaluation (i.e., publication in peer review journals). The selection of an evaluation tool should consider the following factors:

- Type of environment being assessed
- Level of expected risks
- Types of risks
- User population
- Compatibility of tool in assessing task risk factors (i.e., can it assess primary risk factors)
- Economic resources available to support the test

A job analysis tool designed to provide a systematic approach to a holistic understanding of the operator and work environment is the "Arbeitswissenschaftliches Erhebungsverfahren zur Tatigkeitsanalyse" or AET method, translated as the "Ergonomic Job Analysis Technique." This technique considers, measures, or evaluates relevant areas of the following in task performance (Luczak, 1997):

- Work object
- Work resources
- Working environment
- Work task
- Work requirements

Table 6.5 provides details associated with each of these aspects of the evaluation using the AET method.

This is a theoretical approach that can be applied utilizing any number of resources (analysis tools) to perform the assessments in each category and can be useful in assuring a systematic and comprehensive evaluation of the work environment.

JACK ergonomic software is a human-centric modeling software system developed by the Center for Human Modeling and Simulation at the University of Pennsylvania. This product can support research, design, and product development by providing users with a computer-based tool that allows the evaluation of various design and system parameters relevant to biomechanics, ergonomics, and physiology.

TABLE 6.5
Contents of the AET

Part A—Work System Analysis

1. Work objects
  1.1. Material work objects (physical condition, special properties of the material, quality of surfaces, manipulation delicacy, form, size, weight, dangerousness)
  1.2. Energy as work object
  1.3. Information as work object
  1.4. Man, animals, plants as work objects
2. Equipment
  2.1. Working equipment
    2.1.1. Equipment, tools, machinery to change the properties of work objects
    2.1.2. Means of transport
    2.1.3. Controls
  2.2. Other equipment
    2.2.1. Displays, measuring instruments
    2.2.2. Technical aids to support human sense organs
    2.2.3. Work chair, table, room
3. Work environment
  3.1. Physical environment
    3.1.1. Environmental influences
    3.1.2. Dangerousness of work and risk of occupational diseases
  3.2. Organizational and social environment
    3.2.1. Temporal organization of work
    3.2.2. Position in the organization of work sequence
    3.2.3. Hierarchical position in the organization
    3.2.4. Position in the communication system
  3.3. Principles and methods of remuneration
    3.3.1. Principles of remuneration
    3.3.2. Methods of remuneration

Part B—Task Analysis

1. Tasks relating to material work objects
2. Tasks relating to abstract work objects
3. Man-related tasks
4. Number and repetitiveness of tasks

Part C—Job Demand Analysis

1. Demands on perception
  1.1. Mode of perception
    1.1.1. Visual
    1.1.2. Auditory
    1.1.3. Tactile
    1.1.4. Olfactory
    1.1.5. Proprioceptive
  1.2. Absolute/relative evaluation of perceived information
  1.3. Accuracy of perception

(continued)

TABLE 6.5 (continued)
Contents of the AET

Part C—Job Demand Analysis

2. Demands for decision
   2.1. Complexity of decision
   2.2. Pressure of time
   2.3. Required knowledge
3. Demands for response/activity
   3.1. Body postures
   3.2. Static work
   3.3. Heavy muscular work
   3.4. Light muscular work, active light work
   3.5. Strenuousness and frequency of movements

Specifically, this tool allows users to improve product design and workplace tasks through the positioning of biomechanically accurate digital humans of various sizes in virtual environments for evaluation. JACK is a human simulation tool that supports ergonomics and biomechanical analysis by creating animations in which digital humans perform tasks. In this system, biomechanically accurate digital humans of various sizes can be placed in virtual environments, assigned tasks, and analyzed for performance. A result of the detailed models available for evaluation is a realistic condition for in-depth ergonomic assessment (Figure 6.9) (Karwowski and Marras, 2003).

JACK, or other similar products, can improve the ergonomics of product designs and workplace tasks by supporting designers and ergonomists in understanding

FIGURE 6.9   Jack Siemens ergonomic software.

the impact of environmental and physical factors prior to exposing employees to these conditions.

6.7.1 Rapid Upper Limb Assessment Tool

RULA, or rapid upper limb assessment, is a survey method for determining risks associated with upper body postures. It is a screening tool that assesses biomechanical and postural loading on the whole body and focuses on the neck, trunk, and upper limbs. It assesses postural loading at a specific moment in the work cycle. The highest risk posture for analysis may be chosen based on the duration of the posture or the degree of postural deviation. Right or left sides of the body can be assessed, or both sides can be assessed simultaneously. For long work cycles, postures can be evaluated at regular intervals. RULA scores indicate the level of intervention required to reduce MSDs. The following steps are used when scoring and recording the posture:

- Select the stage of the work cycle to be assessed.
- Determine whether the right, left, or both arms will be assessed.
- Score the posture of each limb using the RULA score sheet.
- Review the scoring for accuracy.
- Use tables or software to calculate the grand RULA score.

After the grand score is received, compare the grand RULA score to the action level list to determine the risks. Review the body segment scores for any undesirable postural deviations that may require correction and review possibilities for further ergonomic actions to improve the posture where necessary.

6.7.2 Risk Factors and Risk Prediction

Most of the screening methods described in the previous section involve complex methods to measure the postural or workplace design impacts quantifiably. However often, qualitative measures are used, and a method developed by McCauley-Bell and Badiru in 1993 can be utilized to quantify the risks associated with qualitative and subjective factors in predicting the risks of MSDs. The McCauley-Bell and Badiru model takes the three categories of risk into consideration defined as follows:

- Task factors
- Personal factors
- Organizational factors

The risk level is calculated using a fuzzy linear equation for each category, and then a summative value is calculated using the determined risk level for all three categories. This methodology has been applied in various ergonomic studies. A complete explanation of this risk prediction model can be found in McCauley-Bell and Badiru Part 1 (1994) and McCauley-Bell and Badiru Part 2 (1994).

6.8 GENERAL GUIDELINES FOR WORKPLACES WITH CONTROLS AND DISPLAYS

The information age has led to the design of individual workstations that have numerous and often complex controls and displays. According to Van Coltt and Kinkade (1972), when designing workplaces, some general guidelines can be applied (Sanders and McCormick, 1987):

- *First priority*: primary visual tasks
- *Second priority*: primary controls that interact with primary visual tasks
- *Third priority*: control–display relationships (put controls near associated displays, compatible movement relationships, etc.)
- *Fourth priority*: arrangement of elements to be used in sequence
- *Fifth priority*: convenient location of elements that are used frequently
- *Sixth priority*: consistency with other layouts within the system or in other systems

A more detailed discussion of controls and display can be found in Chapter 9.

6.9 SUMMARY

Workplace design requires the accommodation of employees of different builds and sizes so that everyone is comfortable and can work in the appropriate postures necessary to avoid injury while accessing tools for task performance. Some common principles of workplace design include the provision of comfortable seating, workstation design, equipment selection and environmental factors. An ergonomic approach should be applied to ensure a safe, healthy, and productive workplace.

Case Studies

Ergonomics Case Study:

The Dow Chemical Company's Use of the "Six Sigma" Methodology
May 15, 2004

The Problem: Reducing Musculoskeletal Disorders

Ergonomics-related injuries, including musculoskeletal disorders (MSDs) caused by repetitive strains, continue to be a serious problem for employers. In 2002, ergonomics-related injuries accounted for a third of all workplace injuries involving missed work time, with an average absence of nine days per injury.[1] The resulting worker injury claims and loss of productivity are estimated to cost $13 to $20 million per year for U.S. employers.[2] As computer workstation users spend more and more time at desktops, the risk of MSDs occurring has increased. Yet, as illustrated below, in many companies there are inherent difficulties and concerns associated with addressing this increased ergonomics risk.

For example, Tricia, the Environmental, Health and Safety (EH&S) Leader for the Specialty Chemicals Business of The Dow Chemical Company, wants to reduce MSDs among computer workstation users throughout her business' various divisions and operations. Before she can understand what changes to make in either the workstations or the work practices in those divisions, she must identify the root causes of MSDs among the operators. Although she has some theories, Tricia does not know for sure what factors are causing or contributing to the employees' MSD complaints. Only by knowing the root causes can she implement with confidence controls that would achieve positive results.

Tricia also suspects, but is not sure, that many of the root causes of MSDs are the same across the different operations and divisions in her business. Because of constraints on both her budget and time, Tricia would like to design one basic program that is flexible enough to implement company-wide. She also knows that any reductions achieved under the new program must be sustained over the long term, and she is concerned that over time employees and managers will "backslide" on their commitment to the program and return to their ergonomically risky behaviors.

Fortunately for Tricia, she could refer to a similar project successfully undertaken by the Design and Construction function of The Dow Chemical Company, which is discussed in the case study below. This project, which utilized a problem-solving methodology called "Six Sigma," offered an innovative way to address Tricia's concerns for the development and implementation of a sustainable program to reduce MSDs throughout her business.

The Solution

The Dow Chemical Company's Innovative Use of "Six Sigma"[3]

Avoiding ergonomics-related injuries is an important component of The Dow Chemical Company's ("Dow" or "the Company") overall emphasis on safety and health. Dow is a science and technology company that develops, manufactures, and provides various chemical, plastic, and agricultural products and services for customers in over 180 countries. In 1994, Dow adopted a set of voluntary 10-year EH&S goals to dramatically improve the Company's performance by 2005. These goals call for a reduction in the Company's reportable injury and illness rate by 90% to 0.24%.

In 2000, the company identified an opportunity to improve its injury rate within the Dow Design and Construction business unit. Dow Design and Construction ("DDC") is responsible for managing the design and construction of Dow's facilities worldwide. Because DDC's approximately 1250 workers (including employees and contractors) work primarily at desktop workstations, where they spend the majority of their time working at computer keyboards, they were increasingly susceptible to ergonomics injuries. While the rate of ergonomics-related injuries among the DDC workers was low (only three were reported in 1999), the Company chose to make proactive improvements before ergonomic injuries increased in number or severity.

Dow's EH&S function decided to address ergonomic injuries at DDC using the "Six Sigma" problem solving methodology. Six Sigma is a disciplined, process-oriented approach to problem solving, adopted by Dow and many other companies, which emphasizes the reduction of defects in processes, products and services by applying a four-step improvement methodology. Because Six Sigma emphasizes sustainable results over short-term fixes, Dow has found it particularly useful for EH&S projects. Following the steps prescribed under Six Sigma, Dow developed a Six Sigma project team, which first defined the primary contributing factors to MSDs in the DDC function, and then sought to reduce those factors by 70%. While each of the four steps of the Six Sigma project are outlined below, a more detailed discussion of the Six Sigma methodology appears at the end of this case study.

Step 1: Measure

Once the Six Sigma project team developed its charter and defined its task, it then began by defining the current process. First, the team outlined the sequence of events from workstation assignment to task performance and potential injury. They next identified a series of key variables affecting the process outcome that included:

User attributes (such as daily time at workstations)
User behaviors (including posture, force, and duration of use)
Environmental factors

In this phase of the Six Sigma method, the "defect"—a measurable outcome of the process for which improvement is desired—is defined. While the true "defect" for this process would be the occurrence of an ergonomic injury, there were so few at the start of the project that measuring a statistically significant improvement was going to be difficult. Therefore, the key process variables identified were taken as the "defect," and a goal of 70% improvement (reduction) in the baseline level was set for the project. Scored surveys of DDC workstation users were developed and conducted on the variables identified and used to measure the baseline defect level.

Step 2: Analyze

Accurately identifying the root causes of a problem, which in turn leads to more effective improvements, is an essential function of the Six Sigma methodology. Therefore, the project team next analyzed the collected survey data to determine differences in the workstations, work environments, user training, and behavior at the different DDC sites. The team then identified possible root causes underlying these variables using several of the Six Sigma tools and methodologies, including brainstorming, "fishbone" diagramming, a work performance matrix, and Antecedent-Behavior-Consequence and Balance of Consequences analyses. After developing a list of possible root causes, the team used additional Six Sigma tools and methodologies to identify probable root causes and validate them. For example, one possible root cause identified was a failure of the employee to recognize the importance of ergonomics compliance to his or her personal well-being. This root cause was validated by the employee survey, in which many of the employees expressed an attitude of "it won't happen to me."

Other key root causes validated through this process were the lack of adjustable furniture at some worksites and a lack of "ownership" in personal safety on the part of the employee. The team also determined that ergonomics was not emphasized by DDC to the same extent as other, more immediate, safety issues such as the use of personal protective equipment in hazardous environments.

Step 3: Improve

After determining the most significant root causes through analysis and validation, the project team developed a series of improvements to correct the identified root causes, including both work-related and personal risk factors. Workstation deficiencies were easily addressed by implementing a workstation upgrade plan. Elevating workstation ergonomics to the same level of importance as other personal safety and health issues was a more challenging improvement. However, the team elevated the focus on workstation ergonomics by improving awareness on the part of management and employees and by altering employee behavior and work habits through increased accountability.

The project team developed a novel approach to raising employee awareness by collecting a series of personal testimonials from other employees and posting them on the Company's intranet site. These testimonials were supplemented by more traditional communications, including regular work group safety meetings, training, and increased ergonomics resources. At each facility, the company also designated Ergonomic Focal Points and Ergonomic Contacts, DDC workers who volunteered to receive specialized training and be available as a first point of contact for ergonomic concerns and questions. The team addressed employee behavior by providing feedback to individuals, creating a specific channel for early reporting of discomfort, and developing a health assessment program to address the early warning signs of potential MSDs. Employee personal accountability was addressed by implementing a "Safety First" mentality that stressed ergonomics as a key issue in personal safety and not a separate stand-alone topic.

These improvements are not static, but are a part of an ongoing ergonomics safety and health process. For example, while furniture improvements have been implemented, it is understood that the workstations will continually evolve to meet the employees' changing needs.

Step 4: Control

After the immediate improvements were implemented, the project team developed a long-term control plan designed to sustain the achievements. The control plan took the sequence of events which might contribute to an injury, as outlined in the Measure step, and added a series of performance standards, measures, responsibilities, and contingency plans. For example, in the original sequence, an employee was instructed to attend ergonomics training when starting a job, but there was no control measure to ensure this took place. Under the control plan, the employee is now required to attend the training within 30 days of job assignment, and the designated Ergo Contact at the job site is alerted and follows up with the employee if the employee fails to attend within that timeframe. Each step in the sequence has a similar control, ensuring that the improved process is followed long after the conclusion of the project.

Results of the Project

DDC made immediate improvements in the identified risk factors, which have been reduced 64% since the baseline measurement and by more than 45% overall. These improvements have been well received by the DDC's management and workers, and employees are proactive in addressing discomfort and have a better understanding of the personal benefits of ergonomics. As improvements like these have been repeated throughout the Company, the severity of ergonomics injuries has declined. In 2001, 53% of the Company's ergonomic injuries resulted in lost work time or advanced medical treatment. However, in 2003, only 30% of ergonomic injuries were this severe; the remaining 70% of cases required only first aid or precautionary measures. This result, in turn, has contributed to Dow's 2005 goal of reducing the Company's reportable injury and illness rate by 90% to 0.24%.

Moreover, by virtue of the Six Sigma Methodology's emphasis on long-term control, the project has developed an ongoing process that will help the DDC sustain its immediate results and continue to improve. The positive results of this project have been shared with Tricia and other EH&S managers at other business units, leading to similar projects throughout the company.

Dow believes that using Six Sigma for EH&S projects such as these enables employers to develop program improvements based on measurement and analysis, rather than speculation, resulting in a more cost-efficient and sustainable fix that will yield benefits indefinitely. Rather than undertaking costly trial and error attempts at solutions, the Company was able to identify the root causes of ergonomic injuries with confidence and make improvements to the ergonomics program in a systematic and sustainable way.

Sidebar: Six Sigma Methodology

The Greek letter sigma (σ) is used in mathematics to represent standard deviation, or how much a process varies from its average value. Under the Six Sigma methodology, deficiencies are described in terms of "defects" per million opportunities, with the score of 6s equal to 3.4 defects per million opportunities. Six Sigma uses the following four-step process known as MAIC (Measure, Analyze, Improve, Control) to significantly reduce defects in processes, products, and/or services:

Step 1: Measure—clearly define the process to be improved and the "defect" for the project, and identify a clear and appropriate measure for the "defect."

Step 2: Analyze—determine the root causes of the defect.

Step 3: Improve—develop solutions to address the root causes and validate process improvement.

Step 4: Control—implement a long-term strategy to ensure that the improvements are sustained.

The methodology can be applied to any process that allows the measurement of benefits and improvements in defect reduction, whether in the manufacture of a product, the delivery of a service, the control of costs, or the management of injuries and illnesses.

Dow has adopted the Six Sigma methodology to accelerate the company's improvement in quality and productivity. Dow has expanded the use of the Six Sigma approach to help manage aspects of the Company's operations beyond production and quality, including the safety and health of its workforce. Some of the projects to which Dow has applied the Six Sigma methodology include:

Reduction of repetitive stress injuries
Reduction of motor vehicle accidents
Improved safety for visitors (especially contractors)
Site logistics risk reduction
Off-the-job safety process improvement

These projects have been key components of Dow's 2005 Environmental, Health and Safety Goals, which include reducing Dow's reportable injury and illness rate by 90% to 0.24%.

As the example in our case study illustrates, Dow's Environmental Health and Safety (EHS) function has found the Six Sigma methodology particularly useful in identifying and validating root causes that are hard to discern because of their subjectivity, and in focusing improvements to an ergonomics program in ways that caused measurable improvements. Moreover, since the Six Sigma process includes implementation of controls to ensure that achievements are sustained over a long-term period, the Company expects to realize the benefits of its efforts for years to come.

CASE STUDY REFERENCES

1.   March 2004 U.S. Department of Labor News Release regarding Bureau of Labor Statistics Survey of Occupational Injuries and Illnesses.
2.   "A Critical Review of Epidemiologic Evidence for Work-Related Musculoskeletal Disorders of the Neck, Upper Extremity, and Low Back," The National Institute for Occupational Safety and Health.
3.   Dow Chemical Company: Case study was developed from information provided by Karen Kearns, Industrial Hygiene Specialist, and Mark Spence, Manager, North American Health and Safety Regulatory Affairs, The Dow Chemical Company.

This product was funded under GS 35F 5544H for the US Department of Labor, Occupational Safety and Health Administration. The views expressed herein do not necessarily represent the official position or policy of the U.S. Department of Labor.

EXERCISES

6.1 Explain the history of MTM and how it is applicable to ergonomic task design.
6.2 What is the OSHA perspective on ergonomic workplace design?
6.3 Explain Fitts' Law and its applicability in task design.
6.4 Discuss the principles for good hand tool design.
6.5 Use the AET approach to design the analysis of a real or hypothetical manufacturing task.

REFERENCES

Bernard, T.E. (2011). College of Public Health at the University of South Florida Website, http://personal.health.usf.edu/tbernard/ergotools/index.html (last accessed September 2011).

Brody, A. (1993). Space operations and the human factor. *Aerosp. Am.*, 31(10), 18–21.

Dababneh, A., Lowe, B., Krieg, E., Kong, Y., and Waters, T. (2004). Ergonomics–Achecklist for the ergonomic evalution of nonpowered hand tools. *J. Occup. Environ. Hyg.*, 1, D135–D145, ISSN: 1545-9624 print/1545-9632 online.

Eastman Kodak. (1983). *Ergonomic Design for People at Work—Volume 1*, Van Nostrand Reinhold Company Inc.: New York.

Eastman Kodak. (1986). *Ergonomic Design for People at Work—Volume 2*, Van Nostrand Reinhold Company Inc.: New York.

Fitts, P.M. (1954). The information capacity of the human motor system in controlling the amplitude of movement. *J. Exp. Psychol.*, 47, 381–391.

Hand and Power Tools, Occupational Safety and Health Administration (OSHA) Website. (2011). Hand and Power Tools: Construction Safety and Health Outreach Program, http://www.osha.gov/doc/outreachtraining/htmlfiles/tools.html (last accessed September 2011).

Helander, M. (2006). *A Guide to Human Factors and Ergonomics*, CRC Press: Boca Raton, FL.

Karwowski, W. and Marras, W. S. (2003). *Occupational Ergonomics: Principles of Work Design*, CRC Press: Boca Raton, FL, pp. 29–25.

Kroemer, K.H.E. (1986). Coupling the hand with the handle: An improved notation of touch, grip, and grasp. *Hum. Factors* 28, 337–339.

Kroemer, K.H.E., Kroemer, H.J., and Kroemer-Elbert, K.E. (1994). *Ergonomics: How to Design for Ease and Efficiency*, Prentice Hall: Englewood Cliffs, NJ.

Kroemer, K.H.E. and Grandjean, E. (1997), *Fitting the Task to the Human: A Textbook of Occupational Ergonomics*, 5th edn., Taylor & Francis: London, U.K.

Kroemer, K.H.E., Kroemer, H.J., and Kroemer-Elbert, K.E. (2001). *Ergonomics: How to Design for Ease and Efficiency*, 2nd edn., Prentice-Hall, Inc.: Englewood Cliffs, NJ.

Kromer, K.H.E, Kroemer, H.J., and Kroemer-Elbert, K.E. (2010). *Engineering Physiology: Bases of Human Factors Engineering/Ergonomics*, 4th edn., Springer-Verlag: Berlin, Germany.

Kumar, S. (Ed.) (2008). *Biomechanics in Ergonomics*, 2nd edn., CRC Press: New York.

Luczak, H. (1997). Chapter 12: Task analysis, *Handbook of Human Factors and Ergonomics* (Ed. G. Salvendy), John Wiley & Sons: New York, pp. 340–416.

MacKenzie, I.S. and Buxton, W. (1992). Extending Fitts' Law to two-dimensional tasks, *Proceedings of the CHI '92 Conference on Human Factors in Computing Systems*, Monterey, CA, June 3–7, 1992, ACM: New York.

Marras, S. and Karwowski, W. (2006). *The Occupational Ergonomics Handbook: Fundamentals and Assessment Tools for Occupational Ergonomics*, CRC Press, Taylor & Francis Group: Boca Raton, FL.

Maynard, H.B., Stegemerten, G.J., and Schwab, J.L. (1948). *Methods-Time Measurement*, McGraw-Hill: New York.

McCauley-Bell, P. and Relvini, K. (2005). The human element part of confined space operations, *Proceedings of the Institute of Industrial Annual Conference and Industrial Engineering Research Conference*, Atlanta, GA, May 2005, CD-ROM.

Methods Time Measurement (MTM). (2011). Association for Standards and Research. http://www.mtm.org (last accessed September 2011).

National Institute for Occupational Safety and Health (NIOSH). (1997). *Musculoskeletal Disorders and Workplace Factors—A Critical Review of Epidemiologic Evidence for Work-Related Musculoskeletal Disorders of the Neck, Upper Extremity and Low Back*, Publication No. 97-141, U.S. Department of Health and Human Services: Cincinnati, OH.

National Institute of Occupational Safety and Health (NIOSH). (2010). US Centers for Disease Control Website, last accessed December 2010 (Original publication: NIOSH Publication No. 97-117, adapted from Dave Ridyard (1997)).

Occupational Safety and Health (OSHA). (2011). United States Department of Labor Website, http://www.osha.gov/SLTC/etools/computerworkstations/components_chair.html (last accessed September 2011).

OSHA Instruction CPL 2.78, Directorate of Technical Support. (1991). Appendix A-4., e:\doc\liftech.doc (last accessed August 7, 1991).

OSHA eTools. (2011). Occupational Safety and Health (OSHA), United States Department of Labor Website, http://www.osha.gov/dts/osta/oshasoft/index.html (last accessed September 2011).

Pencil History. (2010). http://www.pencils.com/pencil-history

Petit, T. and Linn, H. (1987). *A Guide to Safety in Confined Spaces*, Publication No. 87-113, DHHS (NIOSH): Washington, DC.

Rekus, J.F. (1994). *Complete Confined Spaces Handbook*, National Safety Council, Lewis Publishers: Boca Raton, FL.

Sanders, M.S. and McCormick, E.J. (1993). *Human Factors in Engineering and Design*, McGraw-Hill: New York. http://catalogs.mhhe.com/mhhe/viewProductDetails.do?isbn=007054901X

Siemens PLM Software. (2010). Siemens Product Lifecycle Management Software Inc., http://www.plm.automation.siemens.com/en_us/products/tecnomatix/assembly_planning/jack/index.shtml?&ku = true&a = 0

Smith, S.S. and Jones, J.H. (1994). A strategy for industrial power hand tool ergonomic research—Design, selection, installation and use in automotive manufacturing, *Proceedings of a NIOSH Workshop*, Cincinnati, OH, January 13–14, 1994.

US Bureau of Labor Statistics (BLS) Website. (2004). Document: CFTB 04/22/2010, TABLE A-9. Fatal occupational injuries by detailed event or exposure, 1996–2004, http://www.bls.gov/iif/oshwc/cfoi/cfoi2004_a-9.pdf

Van Colt, H.P. and Kinkade, R.G. (1972). *Human Engineering Guide to Equipment Design*, American Institute for Research: Washington, DC.

Winn, F., Biersner, R., and Morrissey, S. (1996). Exposure probabilities to ergonomic hazards among miners. *Int. J. Ind. Ergon.*, 18(5–6), 417–422. http://www.sciencedirect.com/science/journal/01698141; http://www.sciencedirect.com/science?_ob=PublicationURL&_hubEid=1-s2.0-S0169814100X00200&_cid=271473&_pubType=JL&view=c&_auth=y&_acct=C000228598&_version=1&_urlVersion=0&_userid=10&md5=86fb8e80eb47186dc65cedf417f555e9 (last accessed December 1996).

WORKPLACE EVALUATION RESOURCE LIST

Posture Evaluations

JACK
Manual Task Risk Assessment Tool (ManTRA) V2.0
Rapid Upper Limb Assessment (RULA)
Rapid Entire Body Assessment (REBA)
Job Strain Index (JSI)
Quick Exposure Checklist (QEC)
Cornell University Body Discomfort Survey 2010 (English—Web form)
Cornell Musculoskeletal Discomfort Questionnaires (English)
Laptop Specific Cornell Musculoskeletal Discomfort Questionnaires (English)
NIOSH Lifting Equation
Liberty Mutual Force Tables (Snook Tables)
Push/pull/carry force calculator

Place Evaluations

USGBC Ergonomics Requirements Worksheets and Survey
Cornell Digital Reading Room Ergonomics Checklist (CDRREC)
OSHA's VDT Checklist
A User-Friendly Computer Workstation Ergonomics Checklist (English)

Source: Cornell University, Ithaca, NY.

7 Work-Related Musculoskeletal Disorders

7.1 LEARNING GOALS

The learning goals of this chapter are to provide an understanding of the origin, nature, and impact of work-related musculoskeletal disorders (WMSDs) in the global community. The student will also learn of the risk factors, evaluation techniques on a general process to establish an ergonomic program focused on reducing the risks of WMSDs in industry.

7.2 KEY TOPICS

- Structures impacted by WMSD
- Types of work-related musculoskeletal disorders
- Stages of work-related musculoskeletal disorders and principles of prevention
- Task-related risk factors
- Personal risk factors
- Screening methods
- Impact on industry
- OSHA ergonomic guideline for WMSD management

7.3 INTRODUCTION AND BACKGROUND

The World Health Organization (WHO), recognizing the impact of "work-related" musculoskeletal diseases, has characterized WMSDs as multifactorial, indicating that a number of risk factors contribute to and exacerbate these maladies (Sauter et al., 1993). The presence of these risk factors produced increases in the occurrence of these injuries, thus making WMSDs an international health concern. These types of injuries of the soft tissues are referred to by many names, including WRMDs, repetitive strain injuries (RSI), repetitive motion injuries (RMI), and cumulative trauma disorders (CTDs). The economic impact of these injuries is felt globally, particularly as organizations begin to develop international partnerships for manufacturing and service roles. Effective guidelines to reduce these risks and associated impacts can benefit the entire global community.

Musculoskeletal disorders have been diagnosed for many years in the medical field. In the nineteenth century, Raynauld's phenomenon, also called dead finger or jackhammer disease, was found to be caused by a lack of blood supply and related to repetitive motions. In 1893, Gray gave explanations of inflammations of the extensor tendons of the thumb in their sheaths after performing extreme exercises.

Long before the Workers' Compensation Act was passed in Great Britain (1906) and CTDs were recognized by the medical community as an insurable diagnosis, workers were victims of the trade they pursued. Since these injuries only manifest themselves after a long period of time, they often went unrecognized. When the demand for labor increased dramatically after the Civil War, the number of children under the age of 15 working in industrial settings skyrocketed to 2 million. Many of these children, who were mostly underweight and malnourished, developed spine curvatures and suffered from stunted growth due to years of industrial labor. Those who worked in coal mines or cotton mills experienced the detrimental effects of breathing in those work environments, like bronchitis and tuberculosis. Additionally, accident rates increased due to the fatigue, as work hours had no legal limit (National Archives, 2011).

Some of the most common CTDs that surfaced were "anthrax from herding and sheep exposure; heavy metal poisoning; a variety of conditions from mining including beat knee, beat elbow and beat hand; inflammation of a variety of joints from mining; and hand difficulties called telegraphist's, writer's and twister's cramp" (Mandel, 2011). In 1929, the state of Ohio official recognized tenosynovitis of the hand as a medical condition, and over the next 25 years, carpel tunnel syndrome (CTS) was fully described in medical books. A full list of upper limb disorders, like "washer woman's sprain," "gamekeeper's thumb," "drummer's palsy," "pipe fitter's thumb," "reedmaker's elbow," "pizza cutter's palsy," and "flute player's hand" were eventually grouped into one of the following medical conditions: tendonitis, epicondylitis, CTS, cubital tunnel syndrome, myalgia, deQuervain's tenosynovitis, or traumatic arthritis (Mandel, 2011).

In 1934, Hammer confirmed that tendons could not resist more than 1500–2000 exertions per hour (Armstrong, 2002). During the 1930s and all the way through the 1950s, reports of inflammation of tendons and tendon sheaths in typists were very common. In the 1940s, both engineers and physicians were more knowledgeable of the impact that the design and operation of work equipment and work schedules had on musculoskeletal disorders. During and after the 1960s, physiological and biomechanical strains of human tissue, particularly of the tendons and their sheaths, revealed that they were indeed associated to repetitive tasks. As a result, several recommendations have been developed for the design and arrangement of workstations, as well as the use of tools and equipment to ultimately alleviate or reduce WMSDs.

7.4 STRUCTURES IMPACTED BY WMSD

Epidemiological studies have shown that in the presence of known risk factors such as force, repetition, and awkward posture, the muscles, joints, tendons, blood vessels, and nerves are at risk for musculoskeletal disorders. Bones are not generally considered to be at risk of these disorders; however, repetitive tasks can be detrimental on the vertebrae of the spinal column.

7.4.1 MUSCLES

Excessively stretching a muscle can lead to a strain. A strain is an injury to a tendon or muscle resulting in swelling and pain. When a group of fibers get torn apart, a more serious injury occurs. Obstruction of blood or nerve supply to the muscle can lead to complete deterioration of the muscle, if the obstruction occurs over an extended period of time.

Collagen fibers in tendons neither stretch nor contract, but if excessively strained, they can be torn. Scar tissue can develop making tendons more prone to repeated injuries and chronic tension. Furthermore, the surfaces on a tendon can become rough, interfering with motion and creating friction along other tissues. When a muscle relaxes or contracts, a tendon produces a gliding movement in its sheath. Some of these movements can be large, for example, when a finger is moved from a fully extended position to a contraction state, the tendon moves 5 cm.

Inflammation of a tendon can occur when there is not enough synovial fluid in the tendon sheath for lubrication and easy gliding. This causes friction between the tendon and its sheath, producing symptoms such as feelings of warmth, tenderness, and pain. These feelings result from an influx of blood, and further compression of the tissue causes more pain. Consequently, any movement of the swollen tendon is hindered and repeated movements can cause more fiber tissues to inflame, creating the possibility of a permanent chronic condition. A bursa is a flat sac that contains synovial fluid and a synovial membrane that facilitate normal movement of bones, muscles, or tendons by reducing friction between them. An overused tendon, especially if roughened, can cause its adjacent bursa to inflame and limits any movement of the tendon, thereby reducing joint mobility.

7.4.2 NERVES

A sprain results from ligament fibers becoming strongly stretched, pulled from the bone, or torn apart. A sprain can stem from either a single injury or repetitive activities producing a cumulative impact. Ligaments that are commonly injured are found in the joint areas such as the knee, ankle, and wrist areas. They can take weeks, or sometimes even months, to completely heal due to their diminished blood supply. A ligament sprain can result in joint instability and can increase the risk of further injury.

Increased pressure within the body occurs when the position of a body part decreases the size of the opening through which nerves run. This is called nerve compression, and it is primarily caused by pressure from ligaments, tendons, muscles, and bones. Also, hard surfaces and sharp edges of workplaces, tools, and equipment can trigger nerve compression. Motor nerves can become impaired, affecting the controlled activity of muscles generated by signals transmitted to the innervated muscle motor nerves. An example is the force or torque required to operate tools, equipment, and other work objects. Impairment of sensory nerves can also occur, making it difficult for information to be transmitted from the sensors in the body to the central nervous system. Symptoms of sensory nerve impairment include sensations of tingling, numbness, or pain. These signals are extremely important because they contain information about the force and pressure applied, positions assumed, and motion experienced; thus, if impaired, it is possible that the operator may not realize certain task activities are increasing the severity of the condition.

7.4.3 ARTERIES AND VESSELS

Vascular compression, often of an artery, occurs when there is a constriction or obstruction of blood flow supply, as well as nutrients and oxygen through tissues, such as muscles, tendons, and ligaments. Vascular compression can result in ischemia, which affects both the duration of muscular activities and the recovery time of a fatigued muscle after performing a task. Another condition, called vasospasm, occurs from vibrations in certain body parts, particularly in the hands and fingers. Vasospasms decrease the diameter of blood vessels and, therefore, impede blood flow to the body areas in need of this blood supply. These areas become blanched, and in the hand this condition is known as the white finger (Raynaud's) phenomenon. Symptoms of vasospasms include intermittent or continual numbness and tingling, with the skin becoming pale and cold, and eventually loss of sensation and control. Vasospasms in the fingers are often a result of vibrations from powered hand tools and equipment such as chain saws, power grinders, pneumatic hammers, and power polishers.

7.5 TYPES OF WORK-RELATED MUSCULOSKELETAL DISORDERS

The frequency of work-related musculoskeletal injuries has led to classification and definition of these conditions in the medical and occupational environments. The classification of the conditions allows the scientific community to understand how to treat the conditions, as well as provides information that engineers can utilize to design processes and equipment to mitigate the risk factors. WMSDs can impact all areas of the body and some of the primary WMSDs experienced in the American industry are discussed in the following section.

7.5.1 BACK INJURIES

The back is the most frequently injured part of the body (22% of 1.7 million injuries) (NSC, Accident Facts, 1990) with overexertion being the most common cause of these injuries. However, many back injuries develop over a long period of time by a repetitive loading of the discs caused by improper lifting methods or other exertions. In fact, 27% of all industrial back injuries are associated with some form of lifting or manual material handling. These injuries are generally repetitive and result after months or years of task performance. Often injuries that appear to be acute are actually the result of long-term impact. The discs of the back vary in size, are round, rubber-like pads filled with thick fluid, which serve as shock absorbers. All the forces that come down the spine compress these discs, as a result of continuous and repetitive squeezing. In some instance discs can rupture and bulge producing pressure on the spinal nerve resulting in back pain.

7.5.2 CARPAL TUNNEL SYNDROME

Perhaps the most widely recognized WMSD of the hand and forearm region is CTS, a condition whereby the median nerve is compressed when passing through the bony carpal tunnel (wrist). The carpal tunnel is comprised of eight carpal bones at the wrist, arranged in two transverse rows of four bones each. The tendons of the forearm muscles pass through this canal to enter the hand and are held down on the anterior side by fascia, called flexor and extensor retinacula, which are tight bands of tissue that protect and restrain the tendons as they pass from the forearm into the hand. If these transverse bands of fascia were not present, the tendons would protrude when the hand is flexed or extended. The early stages of CTS result when there is a decrease in the effective cross section of the tunnel caused by the synovium swelling and the narrowing of the confined space of the carpal tunnel. Subsequently, the median nerve, which accompanies the tendons through the carpal tunnel, is compressed and the resulting condition is CTS.

Early symptoms of CTS include intermittent numbness or tingling and burning sensations in the fingers. More advanced problems involve pain, wasting of the muscles at the base of the thumb, dry or shiny palms, and clumsiness. Many symptoms first occur at night and may be confined to a specific part of the hand. If left untreated, the pain may radiate to the elbows and shoulders (Figure 7.1).

7.5.3 TENDONITIS

Tendonitis, an inflammation of tendon sheaths around a joint, is generally characterized by local tenderness at the point of inflammation and severe pain upon movement of the affected joint. Tendonitis can result from trauma or excessive use of a joint and can afflict the wrist, elbow (where it is often referred to as "tennis elbow"), and shoulder joints. (Figure 7.2).

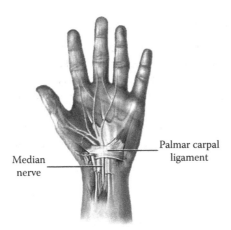

Median
nerve

Palmar carpal
ligament

FIGURE 7.1 Carpal tunnel syndrome.

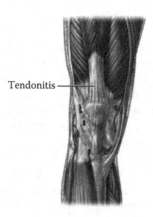

FIGURE 7.2 Tendonitis.

7.5.4 TENOSYNOVITIS

Tenosynovitis is a repetition-induced tendon injury that involves the synovial sheath. The most widely recognized tenosynovitis is DeQuervain's disease. This disorder affects the tendons and sheaths on the side of the wrist and at the base of the thumb.

7.5.5 INTERSECTION SYNDROME AND DEQUERVAIN'S SYNDROME

Intersection syndrome and DeQuervain's syndrome occur in hand-intensive workplaces. These injuries are characterized by chronic inflammation of the tendons and muscles on the sides of the wrist and the base of the thumb. Symptoms of these conditions include pain, tingling, swelling, numbness, and discomfort when moving the thumb. (Figure 7.3).

7.5.6 TRIGGER FINGER

If the tendon sheath of a finger is aggravated, swelling may occur. Sufficient amounts of swelling may result in the tendon becoming locked in the sheath. At this point, if the person attempts to move the finger, the result is a snapping and jerking movement. This condition is called trigger finger. Trigger finger occurs to the individual or multiple fingers and results when the swelling produces a thickening on the tendon that catches as it runs in and out of the sheath. Usually, snapping and clicking in the finger arises with this disorder. These clicks manifest when one bends or straightens the fingers (or thumb). Occasionally, a digit will lock, either fully bent or straightened (See Figure 7.4).

7.5.7 ISCHEMIA

Ischemia is a condition that occurs when blood supply to a tissue is lacking. Symptoms of this disorder include numbness, tingling, and fatigue depending on the degree of

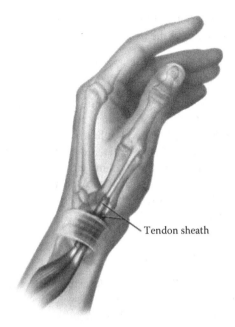

DeQuervain's disease

FIGURE 7.3   DeQuervain's syndrome.

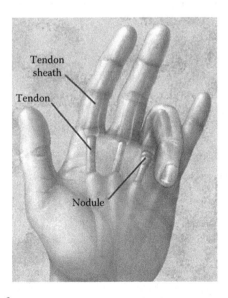

FIGURE 7.4   Trigger finger.

ischemia, or blockage of peripheral blood vessels. A common cause of ischemia is compressive force in the palm of the hand.

7.5.8 VIBRATION SYNDROME

Vibration syndrome is often referred to as white finger, dead finger, or Raynaud's phenomenon. These conditions are sometimes referred to as hand arm vibration (HAV) syndrome. Excessive exposure to vibrating forces and cold temperatures may lead to the development of these disorders. It is characterized by recurrent episodes of finger blanching due to complete closure of the digital arteries. Thermoregulation of fingers during prolonged exposure to cold is recommended, as low temperatures reduce blood flow to the extremities and can exacerbate this condition (See Figure 7.5).

7.5.9 THORACIC OUTLET SYNDROME

Thoracic outlet syndrome (TOS) is a term describing the compression of nerves (brachial plexus) and/or vessels (subclavian artery and vein) to the upper limb. This compression occurs in the region (thoracic outlet) between the neck and the shoulder. The thoracic outlet is bounded by several structures: the anterior and middle scalene muscles, the first rib, the clavicle, and, at a lower point, by the tendon of the pectoralis minor muscle. The existence of this syndrome as a true clinical entity has been questioned, because some practitioners suggest that TOS has been used in error when the treating clinician is short on a diagnosis and unable to explain the patient's complaints. Symptoms of TOS include aching pain in the shoulder or arm, heaviness or easy fatigability of the arm, numbness and tingling of the outside of the arm or especially the fourth and fifth fingers, and finally

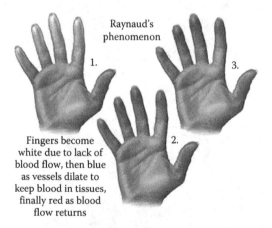

FIGURE 7.5 *Raynaud's Phenomenon:* This condition is characterized by the following stages: (1) Fingers becoming white due to lack of blood flow, then (2) blue due to oxygen consumption, and (3) finally red as blood flow returns.

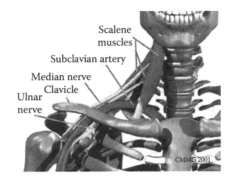

Scalene muscles
Subclavian artery
Median nerve
Clavicle
Ulnar nerve
CMMG 2001

FIGURE 7.6 Thoracic outlet syndrome.

swelling of the hand or arm accompanied by finger stiffness and coolness or pallor of the hand (See Figure 7.6).

7.5.10 GANGLION CYSTS

Ganglion is a Greek word meaning "a knot of tissue." Ganglion cysts are balloon-like sacs, which are filled with a jelly-like material. The maladies are often seen in and around tendons or on the palm of the hand and at the base of the finger. These cysts are not generally painful and with reduction in repetition often leave without treatment (Table 7.1) (See Figure 7.7).

7.6 STAGES OF WMSD AND PRINCIPLES OF PREVENTION

Although *MSDs are cumulative in nature, the recognition of these disorders varies from person to person.* The signs and symptoms of musculoskeletal disorders can appear either slowly or suddenly. Three stages have been established to categorize these disorders gradually:

- Stage 1
  - Workers experience momentary aches and tiredness during normal working hours.
  - Generally, symptoms go away on their own overnight and over days off from work.
  - Work performance is not affected during this stage, but the symptoms can continue for weeks or even months.
- Stage 2
  - Symptoms include tenderness, swelling, weakness, numbness, and pain that begin early in the work shift and do not go away overnight.
  - Workers experiencing stage 2 symptoms may have a difficult time sleeping due to the pain and discomfort.
  - There is a reduction in work performance, specifically repetitive work.
  - Stage 2 symptoms usually last for months.

TABLE 7.1

Summary Table of WMSD

Identified Disorders, Occupational Risk Factors, and Symptoms

Disorders	Occupational Risk Factors	Symptoms
Tendonitis/ tenosynovitis	Repetitive wrist motions Repetitive shoulder motions Sustained hyperextension of arms Prolonged load on shoulders	Pain, weakness, swelling, burning sensation, or dull ache over affected area
Epicondylitis (elbow tendonitis)	Repeated or forceful rotation of the forearm and bending of the wrist at the same time	Same symptoms as tendonitis
Carpal tunnel syndrome	Repetitive wrist motions	Pain, numbness, tingling, burning sensations, wasting of muscles at base of thumb, dry palm
deQuervain's disease	Repetitive hand twisting and forceful gripping	Pain at the base of thumb
Thoracic outlet syndrome	Prolonged shoulder flexion Extending arms above shoulder height Carrying loads on the shoulder	Pain, numbness, swelling of the hands
Tension neck syndrome	Prolonged restricted posture	Pain

Source: Canadian Occupational Safety and Health, http://www.ccohs.ca/oshanswers/diseases/rmirsi.html (accessed April, 2011)

- Stage 3
  - Symptoms during this stage persist even when the person is at complete rest.
  - Frequently, sleep is disturbed and pain is felt even with non-repetitive motions.
  - Work performance is highly affected, even when performing light tasks in daily life.
  - Stage 3 symptoms persist for months or years.

The action to reduce risks should be taken immediately upon recognition of these symptoms. Stage 1 symptoms can be relieved through work changes, such as taking more rest breaks throughout the work shift and task rotation. When experiencing stage 2 or stage 3 symptoms, it is extremely important to completely abandon the task or tasks that are creating the pain or injuries. Additional care may be required including physiotherapy, drugs, and other medical treatments, including surgery. Early detection of a musculoskeletal disorder is essential, as it allows for more efficient treatment and complete recovery. Ergonomically, it is crucial to identify any potentially harmful activities as early as possible in order to mitigate or avoid risks either through work modifications or work redesign.

The identification of WMSDs can sometimes be difficult in the medical and occupational setting due to the cumulative progression of these conditions. A guideline

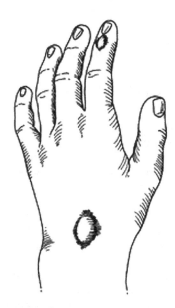

FIGURE 7.7 Ganglion Cyst on hand.

for the minimal clinical criteria for establishing worksite diagnoses has been proposed by Ranney et al. (1995). These criteria provide expected symptoms and examination procedures for baseline evidence of WMSDs (Tables 7.2 and 7.3).

7.7 TASK-RELATED RISK FACTORS

A comprehensive review of epidemiological studies was performed to assess the risk factors associated with WMSDs (*Musculoskeletal Disorders and Workplace Factors*, NIOSH, 1997). The review categorized WMSDs by the body part impacted including (1) neck and neck-shoulder, (2) shoulder, (3) elbow, (4) hand-wrist, and (5) back. The widely accepted task-related risk factors include repetition, force, posture, vibration, temperature extremes, and static posture (NIOSH, 1997). The degree of impact is synergistic when two or more factors exist. The risk factors for each region of the body are described in this review as follows:

1. Neck and neck/shoulder risk factors
   a. *Repetition*—cyclical work activities that involve either repetitive neck movements (i.e., the frequency of different head positions during a cycle), or repeated arm or shoulder motions that generate loads to the neck-shoulder area (i.e., trapezius muscle).
   b. *Force*—forceful exertions involving the upper body that generates loads to the trapezius and neck muscles.
   c. *Posture*—adverse or extreme head or neck postures or static postures of the head and/or neck.

TABLE 7.2
Minimal Clinical Criteria for Establishing Worksite Diagnoses for Work-Related Neuritis

Disorder	Symptoms	Examination
Carpal tunnel syndrome	Numbness and/or tingling in thumb, index, and/or midfinger enhanced symptoms at night	Positive Phalen's test or Tinel's sign with particular writ postures and compression of the median nerve
Scalenus anticus syndrome	Numbness and/or tingling on the preaxial border of the upper lip	Tender scalene muscles with positive Adson's or Wright's test
Cervical neuritis	Pain, numbness, or tingling following a dermatomal pattern in the upper limb	Clinical evidence of intrinsic neck pathology
Lateral antebrachial neuritis	Lateral forearm pain, numbness, and tingling	Tenderness of coracobrachialis origin and reproduction of symptoms on palpation here or by resisted coracobrachialis activity
Pronator syndrome	Pain, numbness, and tingling in the median nerve distribution distal to the elbow	Tenderness of pronator teres or superficial finger flexor muscle, with tingling in the median nerve distribution on resisted activation of same
Cubital tunnel syndrome	Numbness and tingling distal to elbow in ulnar nerve distribution	Tender over ulnar nerve with positive Tinel's sign and/or elbow flexion test
Ulnar tunnel syndrome	Numbness and tingling in ulnar nerve distribution in the hand distal to the wrist	Positive Tinel's sign over the ulnar nerve at the wrist
Wartenberg's syndrome	Numbness and/or tingling in distribution of the superficial radial nerve	Positive Tinel's sign on tapping over the radial sensory nerve
Digital neuritis	Numbness or tingling in the fingers	Positive Tinel's sign on tapping over digital nerves

Source: Adapted from Ranney, D. et al., *Ergonomics*, 38(7), 1408, 1995. Reprinted with permission.

2. Shoulder risk factors
   a. *Repetition*—task activities that involve cyclical flexion, extension, abduction, or rotation of the shoulder joint.
   b. *Posture*—when the arm is flexed, abducted, or extended, such that the angle between the torso and the upper arm increases, and the greater this angle, the greater the impact of this risk factor.
   c. *Force*—strenuous work involving shoulder abduction, flexion, extension, or rotation to exert force thus generating loads in the shoulder.
   d. *Vibration*—low- or high-frequency vibration generally as a result of hand tools.

TABLE 7.3
Minimal Clinical Criteria for Establishing Worksite Diagnoses for Work-Related Muscle or Tendon Disorders

Disorder	Symptoms	Examination
Neck myalgia	Pain one of both sides of neck increase by neck movement	Tender over paravertebral neck muscles
Trapezius myalgia	Pain on top of shoulder increased by shoulder elevation	Tender top of shoulder or medial border of scapula
Scapulothoracic pain syndrome	Pain in scapular region increased by scapular movement	Tender over rib angles 2, 3, 4, 5, and/or 6
Rotator cuff tendonitis	Pain in deltoid area or front of shoulder increased by glenohumeral movement	Rotator cuff tenderness
Triceps tendinitis	Elbow pain increased by elbow movement	Tender triceps tendon
Arm myalgia	Pain in muscle(s) of the arm	Tenderness in a specific muscle of the arm
Epicondylitis/tendonitis	Pain localized to lateral or medial aspect of elbow	Tenderness of lateral or medial epicondyle localized to this area or to soft tissues attached for a distance of 1.5 cm
Forearm myalgia	Pain in the proximal half of the forearm (extensor or flexor aspect)	Tenderness in a specific muscle in the proximal half of the forearm (extensor or flexor aspect) more than 1.5 cm distal to the condyle
Wrist tendonitis	Pain on the extensor or flexor surface for the wrist	Tenderness is localized to specific tendons and is not found over bony prominences
Extensor finger tendonitis	Pain on the extensor surface of the hand	Tenderness is localized to specific tendons and is not found over bony prominences
Flexor finger tendonitis	Pain on the flexor aspect of the hand of distal forearm	Pain on resisted finger flexion localized to area of tendon
Tenosynovitis (finger/thumb)	Clicking or catching of affected digit on movement. There may be pain or a lump in the palm	Demonstration of these complaints, tenderness anterior to metacarpal of affected digit
Tenosynovitis, deQuervain's	Pain on the radial aspect of wrist	Tenderness over first tendon compartment and positive Finkelstein's test
Intrinsic hand myalgia	Pain in muscles of the hand	Tenderness in a specific muscle in the hand

3. Elbow risk factors
   a. *Posture*—activities that require repeated pronation, supination, flexion, or extension of the wrist, either singly or in combination with extension and flexion of the elbow.
   b. *Force*—strenuous activities involving the forearm extensors or flexors, which can generate loads to the elbow–forearm region.
   c. *Repetition*—(1) cyclical flexion and extension of the elbow or (2) cyclical pronation, supination, extension, and flexion of the wrist that generates loads to the elbow–forearm region.
4. Hand-wrist risk factors
   a. *Repetition*—cyclical or repetitive work activities that involve either repetitive hand-finger or wrist movements (i.e., hand gripping) or wrist extension-flexion, ulnar-radial deviation, and supination or pronation. Frequent repetitions have been defined as a cycle time <30 s or 50% of the task cycle spent performing the same activity (Silverstein et al., 1987).
   b. *Posture*—deviations from the natural posture of the hand, wrist and/ or fingers-wrist flexion or extension, ulnar or radial deviation full hand grip, and pinch grip.
   c. *Force*—forceful exertions performed by the hand, with or without a hand tool, during manipulative task activities.
   d. *Hand arm vibration (HAV)*—manual work involving vibrating power hand tools. HAV is the transfer of vibration from a tool to a worker's hand and arm. The level of HAV is a function of the acceleration level of the tool when grasped by the worker when in use.
   e. *Cold environments*—compromise muscle efficiency and may cause vascular and neurological damage. Workers with cold hands may exert more force than necessary, affecting muscles, soft tissues, and joints. Also, cold environments may require gloves that have been shown to impact sensation thus leading to additional force exertion. This can lead to a more rapid onset of fatigue and to the development of disorders.
5. Lower back risk factors
   a. *Heavy physical work*—task activities with high-energy demands requiring some measure of physical strength and can include manual material handling tasks, heavy dynamic, or intense work. Heavy work can also be tasks that impose large compressive forces on the spine (Marras et al., 1995).
   b. *Lifting/forceful movement*—the physical stress that results from work done in transferring objects from one plane to another, as well as the effects of forceful movements such as pulling, pushing, or other efforts.
   c. *Awkward posture*—non-neutral trunk postures (related to bending and twisting) in extreme positions or at extreme angles.
      i. Bending is defined as flexion of the trunk, usually in the forward or lateral direction.
      ii. Twisting refers to trunk rotation or torsion.
      iii. Kneeling, squatting, and stooping are also considered awkward postures.

   d. *Whole body vibration*—WBV refers to mechanical energy oscillations
      that are transferred to the body as a whole (in contrast to specific body
      regions), usually through a supporting system such as a seat or platform.
      Typical exposures include driving automobiles and trucks, and operat-
      ing industrial vehicles.
   e. *Static work posture*—isometric positions where very little movement
      occurs, along with cramped or inactive postures that cause static load-
      ing on the muscles. This includes prolonged standing or sitting and sed-
      entary work (NIOSH, 1997).

7.8   PERSONAL RISK FACTORS

The field of ergonomics does not attempt to screen workers for elimination as potential employees. However, on the contrary, Ergonomics is focused on accomodating
a variety of workers. The recognition of personal risk factors can be useful in providing training, administrative controls, and awareness. Personal or individual risk
factors can impact the likelihood for occurrence of a WMSD (McCauley-Bell and
Badiru, 1996). These factors vary depending on multiple factors inculding age, gender, smoking, physical activity, strength, anthropometry and previous WMSD, and
degenerative joint diseases.

7.8.1   AGE

Although WMSD can impact workers at any age, musculoskeletal impairments, particularly of the back, are among the most prevalent occupational problems of middle
aged and older (Buckwalter et al., 1993) workers. Nonetheless, age groups with the
highest rates of compensable back pain and strains are the 20–24 age group for men,
and the 30–34 age group for women (this may be due to an abundance of employees
in these age groups). In addition to decreases in musculoskeletal function due to the
development of age-related degenerative disorders (i.e., arthritis), loss of muscle fiber
and degradation of tissue strength with increasing age may increase the likelihood
and severity of soft tissue damage from a given insult.

7.8.2   GENDER

According to a recent U.S. Government report, women are three times more likely
to have CTS than men (Women.gov, 2011). Women also deal with strong hormonal
changes during pregnancy and menopause that make them more likely to suffer from
WMSD, due to increased fluid retention and other physiological conditions. Other
reasons for the increased presence of WMSDs in women may be attributed to differences in muscular strength, anthropometry, or hormonal issues. Generally, women
are at higher risk of the CTS between the ages of 45 and 54. Then, the risk increases
for both men and women as they age. Some studies have found a higher prevalence
of some WMSDs in women (Bernard et al., 1994; Chiang et al., 1993; Hales et al.,
1994; Johansson, 1994), but the fact that more women are employed in hand-intensive jobs may account for the greater number of reported work-related MSDs among

women. Likewise, Byström et al. (1995) reported that men were more likely to have DeQuervain's disease than women and attributed this to more frequent use of power hand tools. Whether the gender difference seen with WMSDs in some studies is due to physiological differences or differences in exposure is not fully understood. To differentiate the effect of work risk factors from potential effects that might be attributable to biological differences, researchers must study jobs that men and women perform relatively equally.

7.8.3 PHYSICAL ACTIVITY

Studies on physical fitness level as a risk factor for WMSDs have produced mixed results. Physical activity may cause injury. However, the lack of physical activity may increase susceptibility to injury, and after injury, the threshold for further injury is reduced. In construction workers, more frequent leisure time was related to healthy lower backs (Holmström et al., 1991) and severe low-back pain was related to less leisure time activity (Holmström et al., 1992). On the other hand, some standard treatment regimes have found that musculoskeletal symptoms are often relieved by physical activity. National Institute for Occupational Safety and Health (NIOSH, 1991) stated that people with high aerobic capacity may be more fit for jobs that require high oxygen uptake, but will not necessarily be more fit for jobs that require high static and dynamic strengths and vice versa.

7.8.4 STRENGTH

Epidemiologic evidence exists for the relationship between back injury and weak back strength in job tasks. Chaffin and Park (1973) found a substantial increase in back injury rates in subjects performing jobs requiring strength that was greater than or equal to their isometric strength-test values. The risk was three times greater in weaker subjects. In a second longitudinal study, Chaffin et al. (1977) evaluated the risk of back injuries and strength; this study found the risk to be three times greater in weaker subjects. Other studies have not found the same relationship with physical strength. Two prospective studies of low-back pain reports (or claims) of large populations of blue collar workers (Battie et al., 1989; Leino, 1987) failed to demonstrate that stronger workers (defined by isometric lifting strength) are at lower risk for low-back pain claims or episodes.

7.8.5 ANTHROPOMETRY

Weight, height, body mass index (BMI) (a ratio of weight to height squared), and obesity have all been identified in studies as potential risk factors for certain WMSDs, particularly CTS and lumbar disc herniation. Vessey et al. (1990) found that the risk for CTS among obese women was double that of slender women. The relationship of CTS and BMI has been suggested to be related to increased fatty tissue within the carpal canal or to increased hydrostatic pressure throughout the carpal canal in obese persons compared with slender persons (Werner, 1994). Carpal tunnel canal size and wrist size has been suggested as a risk factor for CTS; however, some studies have linked both small and large canal areas to CTS (Bleecker, et al., 1985; Winn and Habes, 1990). Studies on anthropometric data are conflicting, but in general indicate that there is no strong correlation between stature, body weight, body build, and low back pain. The general conclusion is that obesity seems to play a small but significant role in the occurrence of CTS.

7.8.6 SMOKING

Several studies have presented evidence that smoking is associated with low-back pain, sciatica, or intervertebral herniated disc (Finkelstein, 1995; Frymoyer et al., 1983; Kelsey et al., 1984; Owen and Damron, 1984; Svensson and Anderson, 1983); whereas in others, the relationship was negative (Frymoyer, 1993; Hildebrandt, 1987; Kelsey et al., 1990; Riihimäki et al., 1989b). Boshuizen et al. (1993) found a relationship between smoking and back pain only in those occupations that required physical exertion. In this study, smoking was more clearly related to pain in the extremities than to pain in the neck or the back. Deyo and Bass (1989) noted that the prevalence of back pain increased with an increase in packs of cigarettes smoked in a year, with the heaviest smoking level. Several explanations for the relationship have been proposed. One hypothesis is that back pain is caused by coughing from smoking. Coughing increases the abdominal pressure and intradiscal pressure, thereby producing strain on the spine. Several studies have observed this relationship (Deyo and Bass, 1989; Frymoyer et al., 1980; Troup et al., 1987). Other theories include nicotine-induced diminished blood flow to vulnerable tissues (Frymoyer et al., 1983), and smoking-induced diminished mineral content of bone causing microfractures (Svensson and Andersson, 1983).

7.9 SCREENING METHODS AND DIAGNOSIS OF WORK-RELATED MUSCULOSKELETAL DISORDERS

Several methods have been developed to both diagnose and treat musculoskeletal disorders. These methods range from simple checklist (see Table 7.4) to electronic equipment and software-based tools. In the Occupational Safety and Health Administration (OSHA) checklist, a rating of more than 5 suggests a problem job (Karwowski and Marras, 1997).

Job survey methods such as checklist can be very useful in establishing a general understanding of the work environment. In addition to checklists, this information may be obtained prior to formal ergonomic assessments by conducting facility walkthroughs, interviews (with workers and supervisors), and having group discussions. A summary of these proactive job survey methods has been proposed by Karwoski and Marras (1997) and is provided in Table 7.5 (Karwowski and Marras, 1997).

Likewise, surveillance techniques can also be used to assess risk levels. Surveillance approaches may be either passive or active. Passive surveillance methods are used to evaluate information that already exists in an organization where it has usually been obtained for other purposes (i.e., injury statistics, quality control data). On the other hand, active surveillance is the collection of data specifically for the given purposes

TABLE 7.4

Ergonomic Measures to Control Common WMSDs

Disorder	Avoid in General	Avoid in Particular	Recommendation	Design Issues
Carpal tunnel syndrome	Rapid, often repeated finger movements, wrist deviation	Dorsal and palmar flexion, pinch grip, vibration between 10 and 60 Hz		Workplace design
Cubital tunnel syndrome	Resting forearm on sharp edge or hard surface			Workplace design
deQuervain's syndrome	Combined forceful gripping and hard twisting			Workplace design
Epicondylitis	"Bad tennis backhand"	Dorsiflexion, pronation		Workplace design
Pronator syndrome	Forearm pronation	Rapid and forceful pronation, strong elbow and wrist flexion	Use large muscles but infrequently and for short time	Design of work object
Shoulder tendonitis, rotator cuff syndrome	Arm elevation	Arm abduction, elbow elevation	Let wrists be in line with the forearm	Design of job task
Tendonitis	Often repeated movements, particularly with force exertion; hard surface in contact with skin; vibrations	Frequent motion of digits, wrists, forearm shoulder	Let shoulder and upper arm be relaxed	Design of hand tools ("bend tool, not the wrist")
Tenosynovitis, deQuervain's syndrome, ganglion	Finger flexion, wrist deviation	Ulnar deviation dorsal and palmar flexion, radial deviation with firm grip	Let forearms be horizontal or more declined	Design for round corners, use pad
Thoracic outlet syndrome	Arm elevation, carrying loads	Shoulder flexion, arm hyperextension		Design of work object placement
Trigger finger or thumb	Digit flexion	Flexion of distal phalanx alone		Workplace design
Ulnar artery aneurism	Pounding and pushing with heel of the hand			Workplace design

(continued)

TABLE 7.4 (continued)
Ergonomic Measures to Control Common WMSDs

Disorder	Avoid in General	Avoid in Particular	Recommendation	Design Issues
Ulnar nerve entrapment	Wrist flexion and extension	Wrist flexion and extension, pressure of hypothenar eminence		Workplace design
White finger, vibration syndrome	Vibrations, tight grip, cold exposure	Vibrations between 40 and 125 Hz		Workplace design
Neck tension syndrome	Static head posture	Prolonged static head-neck posture	Alternate head-neck postures	Workplace design

Source: Adapted from Kroemer, K. et al., *Ergonomics, How to Design for Ease and Efficiency*, Prentice Hall, Englewood Cliffs, NJ, 1994. Reprinted with permission.

(i.e., ergonomic evaluation of a workplace). Table 7.6 provides examples of surveillance tools/techniques for WMSDs (Karwowski and Marras, 1997).

Actual physical screening approaches can also be used to assess the presence of WMSDs. A few examples of these methods are discussed in the following text.

7.9.1 PHALEN'S TEST

George S. Phalen, an American hand surgeon, studied patients with CTS for over 17 years. Phalen recognized that Tinel's sign could be used to diagnose CTS, described it as "a tingling sensation radiating out into the hand, which is obtained by light percussion over the median nerve at the wrist" (Urbano, 2000). Additionally, Phalen developed a wrist flexing test to diagnose CTS. To perform the Phalen's test, the patient should place their elbows on a table, placing the dorsal surfaces of the hands against each other for approximately 3 min. The patient should perform this maneuver with the wrists falling freely into their maximum flexion, without forcing the hands into flexion (see Figure 7.8). Patients who have CTS will experience tingling or numbness after 1–2 min, whereas a healthy patient without CTS can perform the test for 10 or more minutes before experiencing tingling or numbness (Urbano, 2000).

7.9.2 DURKAN TEST OR CARPAL COMPRESSION TEST

In 1991, John A. Durkan, an American orthopedic surgeon, developed the carpal compression test. In a study of 31 patients (46 hands) with CTS, he found that this compression test was more sensitive than Tinel's or Phalen's tests (Durkan, 1991). The carpal compression test involves directly compressing the median nerve using a rubber atomizer-bulb connected to a pressure manometer from a

TABLE 7.5

A List of Proactive Job Survey Methods for Control of CTDs

Job Survey	Description
Facility walk-throughs	A facility walk-through consists of using firsthand observations to identify exposures to recognized CTD risk factors. Knowledge of the processes, facilities, and schedules is used to conduct observations of representative activities and workers. This method may be effective at identifying the presence of some risk factors, particularly those that are most obvious. However, a single walkthrough or observation may not be sufficient to determine the magnitude, frequency, and duration of exposure to risk factors. The time required to perform a walk-through depends on the work cycle time and variability of the tasks performed, with longer work cycles and jobs with greater task variability needing more observation time. The method can be enhanced by combining it with formal or informal worker and supervisor interviews and/or risk factor checklists.
Worker and supervisor interviews	A worker and supervisor interview consists of asking questions regarding job/ task attributes and associated risk factor exposures. The method relies on the firsthand observations and experiences of worker and supervisor. Such interviews may be formal (written or recorded) or informal (verbal). This method may be effective at identifying the presence of risk factors and associated job attributes not easily garnered by a facility walk-through, particularly risk factor exposure duration, recovery breaks, infrequently performed tasks, or multiple methods at performing tasks. However, the method is subjective. The method is also performed as part of a more comprehensive job analysis.
Risk factor checklists	Checklists consist of a formal procedure in which CTD risk factors are enumerated on a list and checked off by the analyst through firsthand observation of a specific or representative task(s) or job(s). Checklists may have high sensitivity in identifying the presence of some risk factors; however, they may have low specificity. Checklists must be adapted and validated for each industry or occupation. Using checklists requires caution and familiarity with the job, task, or process.
Team problem solving processes	Team problem-solving processes integrate the firsthand observations and experiences of several workers and supervisors regarding job or task attributes and associated risk factor exposures and attempt to develop, test, and implement plausible solutions. Any of the aforementioned surveys tools may be incorporated into such a process.

Source: After ANSI (1996).

sphygmomanometer (see Figures 7.9 and 7.10). This direct compression uses a pressure of 150 mm of mercury for 30 s. The occurrence of pain or paresthesia (tingling) indicates the presence of CTS. Durkan also identified an alternate method of performing the compression test by having the examiner apply even pressure with both thumbs to the median nerve in the carpal tunnel, as shown in Figures 7.9 and 7.10 (Durkan, 1991).

TABLE 7.6
Examples of Tools for WMSD Surveillance

Surveillance Approach	Methods of Surveillance	
	Passive	Active
Workplace risk factors (associated with WMSDs)	• Company dispensary logs • Insurance records • Workers' compensation records • Health (WMSDs) • Accident reports • Transfer requests • Absentee records • Grievances	• Checklists • Questionnaires • Interviews • Physical exams
	• Not really used for WMSD risk factor yet[a]	• Checklists • Questionnaires • Job analysis

Source: Adapted from Kuorinka, I. and Forcier, L. (Eds.), *Work Related Musculoskeletal Disorders (WMSDs): A Reference Book for Prevention*, Taylor & Francis, Boca Raton, FL, 1995. Reprinted with permission.

[a] The use of surrogate measures for exposure (e.g., job title or firm's department) could be viewed as "passive surveillance."

7.9.3 VIBROMETRY TESTING

Another method for the diagnosis of carpal tunnel includes vibrometry testing. With this technique, the middle finger is placed on a vibrating stylus. While the evaluator manipulates vibration by altering the frequencies, the patient indicates whether or not they can detect the stylus vibrating. In theory, those patients with CTS will be less sensitive to vibration. However, the effectiveness of vibrometry testing is debated with some studies able to successfully identify CTS (Jetzer, 1991; Neese and

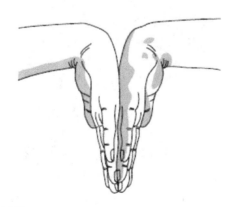

FIGURE 7.8 PHALENS test.

FIGURE 7.9   DURKAN test. (From Durkan, J.A., *J. Bone Joint Surg.*, 73(4), 535, 1991.)

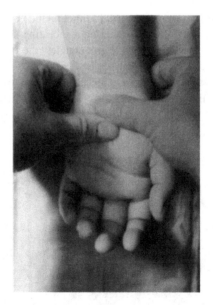

FIGURE 7.10   Alternate DURKAN test. (From Durkan, J.A., *J. Bone Joint Surg.*, 73(4), 535, 1991.)

Konz, 1993), while others show vibrometry testing to be inconclusive (Werner et al., 1994; White et al., 1994).

7.9.4 NERVEPACE ELECTRONEUROMETER DEVICE

Another method for diagnosing carpal tunnel includes the use of the Nervepace Electroneurometer to test motor nerve conduction. Electrodes for surface stimulation are placed on the median nerve 3 cm proximal to the distal wrist flexor crease, while recording electrodes are placed on the muscles of the hand. The evaluator then adjusts the stimulus applied to the median nerve until a motor response is detected. The device records the latency between the stimulus and the response times, which the evaluator can use to determine the presence of CTS. However, studies have shown that the device can be made ineffective due to skin thickness (callous), peripheral neuropathy, or severe CTS. In addition, the American Association of Electrodiagnostic Medicine deemed the Nervepace Electroneurometer as "flawed," "experimental," and "not an effective substitute for standard electrodiagnostic studies in clinical evaluation of patients with suspected CTS" (David and Chaudhry, 2003; Pransky et al., 1997).

7.9.5 TINEL'S SIGN

Jules Tinel, a French neurologist, developed Tinel's sign in 1915. He noted that after an injury, tapping of the median nerve resulted in a tingling sensation (paresthesia) in the first three and a half digits (Figure 7.11). Tinel's sign was not originally associated with the CTS; it was not until 1957 that George Phalen recognized that Tinel's sign could be used to diagnose the CTS (Urbano, 2000).

7.10 IMPACT ON INDUSTRY

The U.S. Bureau of Labor Statistics (BLS) publishes detailed information about occupational injuries, illnesses, and fatalities from private industry. Results of the 2004 survey revealed 1,259,320 nonfatal cases involving days away from work (141.3 cases per 10,000 workers), down from 2003, in which 1,315,920 total cases (150.0 per 10,000 workers) were reported (Table 7.7, U.S. Bureau of Labor Statistics, 2004). These cases included WMSDs. BLS publishes detailed characteristics for WMSD cases that resulted in at least one lost day from work. For example, the number of WMSD cases and incidence rates are reported by sex, age, occupation, industry, and several other characteristics. The BLS data in Table 7.7 reveals that, since 2002, the number and incidence rates of WMSD cases declined. The reason for the decline has been attributed to changes in task design, ergonomic interventions, technology, and greater awareness. According the BLS, these numbers have continued to decrease. There were 335,390 WMSDs in 2007. These accounted for 29% of all workplace injuries resulting in time away from work. In 2006, there were 357,160 cases of WMSDs, making up 30% of all workplace injuries requiring time away from work. It is worth pointing out that these numbers have been reduced; however, there is still a need to incorporate ergonomic design that reduces the risk of these injuries (BLS, 2005).

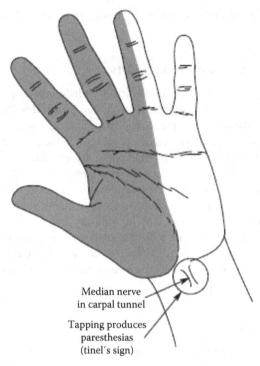

Median nerve
in carpal tunnel

Tapping produces
paresthesias
(tinel's sign)

FIGURE 7.11   Tinel's sign.

Additional research strides are needed to develop predictive models for the development of WMSDs based on the presence of occupational and nonoccupational risk factors. The aforementioned statistics demonstrate why WMSDs continue to be a primary concern of employers. Companies are concerned with the costs associated with WMSDs; therefore, they are eagerly pursuing ways to identify, prevent, and control WMSDs before they become problematic.

Occupations in the American industry that are most prone to WMSD include manual material handling, nursing aides and orderlies, computer operators, ticket agents, and emergency medical personnel. Figure 7.12 illustrates the incidence rates for WMSDs in 2007 in these industries.

Although laborers and manual material handlers are the largest population among WMSD cases, the most heavily affected occupations are nursing aids, orderlies, and attendants with 252 cases per 10,000 workers. Nursing homes and personal care facilities have one of the highest rates of injury and illness among industries for which nationwide lost workday injury and illness (LWDII) rates were calculated for calendar year 2002. According to the BLS, nursing and personal care facilities experienced an average LWDII rate of 7.6, despite the availability of feasible controls that have been identified to address hazards within this industry. This is more than double the LWDII rate of 2.8 for private industry as a whole. These injuries in this industry are growing, as BLS statistics in 2007 reveal that nursing aides, orderlies, and attendants had a WMSD rate more than seven times the national average (U.S.

TABLE 7.7

Number and Incidence Rates[a] of Nonfatal Occupational Injuries and Illnesses with Days away from Work[b] Involving Musculoskeletal Disorders[c], 2004

	WMSD Cases with Days away from Work			
Year	Total Cases	Incidence Rate	Median Days	Relative Standard Error
2002	487,915	55.3	—	—
2003	435,180	49.6	10	0.8
2004	402,700	45.2	10	0.8

Source: U.S. Bureau of Labor Statistics, 2005.

[a] Incidence rates represent the number of injuries and illnesses per 10,000 full-time workers and were calculated as defined in the following: $(N/EH) \times 20,000,000$.

where

N is the number of injuries and illnesses

EH is the total hours worked by all employees during the calendar year

$20,000,000 =$ base for 10,000 full-time equivalent workers (working 40 h/week, 50 weeks/year)

[b] Days away from work cases include those that result in days away from work with or without job transfer or restriction.

[c] Includes cases where the nature of injury is sprains, strains, tears; back pain, hurt back; soreness, pain, hurt, except back; CTS; hernia; or musculoskeletal system and connective tissue diseases and disorders and when the event or exposure leading to the injury or illness is bodily reaction/bending, climbing, crawling, reaching, twisting; overexertion; or repetition. Cases of Raynaud's phenomenon, tarsal tunnel syndrome, and herniated spinal discs are not included. Although these cases may be considered MSDs, the survey classifies these cases in categories that also include non-MSD cases.

Bureau of Labor Statistics, 2008). In response to these growing statistics, NIOSH researchers compiled guidelines for nursing home staff to move patients (Safe Lifting and Movement of Nursing Home Residents, 2006). This document provides guidance on handling of patients in a manner that is safe for them and ergonomically sound for the worker (Figure 7.13).

The National Coalition of Ergonomics estimates that WMSDs result in $15–$20 billion annually, just in worker's compensation costs (Jeffress, 1999). Similarly, the NIOSH estimates costs between $13 and $20 billion annually (Rosenstock, 1997). The Bone and Joint Decade, an international movement supported by the United Nations and the World Health Organization, estimates that the total cost of musculoskeletal diseases in the United States, including treatment and lost wages, was $849 billion (United States Bone and Joint Decade, 2008) further illustrating the international significance of these issues.

WMSDs represent the most important work-related health problems in Europe. In 2003, for every 100,000 workers, there were 32 new WMSD cases. This means that WMSDs account for 50% of all new cases of occupational disorders in the EU (2003 data), and 50% of all work-related health problems in Europe (1999). Moreover, the number of WMSD cases is increasing over time. In 2001, around 19 workers in every 100,000 suffered from a new WMSD. This number had almost doubled by 2003 (32 per 100,000 workers) (OSHA EU, 2011) (Table 7.8).

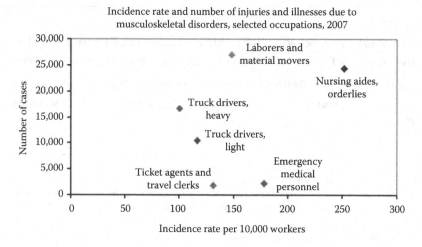

FIGURE 7.12 Incidence rate and number of injuries by industry.

As previously stated, tasks that involve the use of computer keyboard, mouse, and video display terminals (VDT) are also areas of serious concern. In some cases, these jobs are considered a breeding ground for occupational WMSDs. The World Health Organisation (WHO) concluded that "musculoskeletal discomfort was commonplace during work with VDTs" and that "injury from repeated stress ... is possible" (Sauter et al., 1993). Understanding the global market associated with current

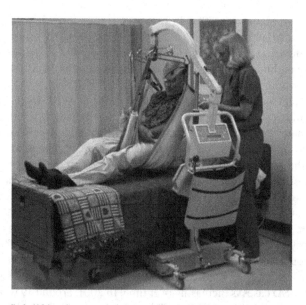

FIGURE 7.13 Safe lifting in a nursing home. (From OSHA website, accessed February 18, 2011.)

TABLE 7.8
Sectors with Highest WMSD in Europe

Sectors with the Highest Relative Number of New MSD Cases (per 100,000 Workers)		Sectors with the Highest Number of Work-Related MSD Complaints (per 100,000 Workers)	
Mining and quarrying	960	Health and social work	4283
Manufacturing	81	Transport and communication	3160
Construction	50	Construction	3158
Total population	32	Total population	2645

organizations, it is imperative to acknowledge and address this problem on an international level. An epidemic of WMSDs affecting VDT operators swept Australia in the last decade. In fact, prevalence rates of almost 35% were recorded in some Australian organizations (Sauter et al., 1991). The high-tech, Japanese industry is also experiencing an adverse impact due to the occurrence of WMSDs. Japanese organizations have reported surges in the number of occupational-related musculoskeletal injuries (Sauter et al., 1991). This further illustrates the need to have comprehensive guidelines for mitigating the risks of work-related musculoskeletal injuries internationally.

WMSDs are clearly a global occupational issue and less than a decade ago were considered the most important work-related health problem in Europe (OSHA EU, 2003). In 2003, for every 100,000 workers, there were 32 new WMSD cases, resulting in WMSDs accounting for 50% of all new cases of occupational disorders in the EU (2003 data), and 50% of all work-related health problems in Europe (1999 data). Efforts have been made to mitigate the impact of these risk factors (OSHA EU, 2003) through the implementation of screening methods, education, and improved workplace design. Ergonomic measures to control common WMSDs include workplace redesign, utilization of ergonomically designed, training hand tools, and other methods to minimize the occurrence of the risk factors. Table 7.4 provides a few ergonomic measures to control common WMSDS (Kroemer et al., 1994).

7.11 PROCEDURE FOR WORKPLACE ANALYSIS AND DESIGN

Job analysis, risk factor assessment, and task design should be conducted to identify potential work-related risks and develop engineering controls, administrative controls, and personal protective resources to mitigate the likelihood of injuries. According to American National Standards Institute (ANSI, 1995), this can be accomplished with the following steps (Karwowski and Marras, 1997):

- Collect pertinent information for all jobs and associated work methods.
- Interview a representative sample of affected workers.
- Breakdown a job into tasks or elements.
- Description of the component actions of each task or element.

- Measurement and qualification or quantification of WMSDs (where possible).
- Identification of risk factors for each task or element.
- Identification of the problems contributing to the risk factors.
- Summary of the problem areas and needs for intervention for all jobs and associated new work methods.

These steps can be executed utilizing any combination of scientifically based assessment techniques including checklists, surveys, electronic measurement equipment, software tools, and analysis approaches.

7.12   ERGONOMIC PROGRAM FOR WMSD

In 1999, OSHA culminated years of scientific study into a proposed ergonomic standard to address the frequency and severity of WMSD in the U.S. occupational environment. The standard was implemented but almost immediately repealed with the transition of leadership in the U.S. presidency. Despite the repealing of the legislation, the guidance offered in the standard serves as a guideline for industries to develop programs to reduce and manage WMSD.

The primary elements of the former OSHA Ergonomics Legislation were developed using as guidelines comprehensive organizational involvement. This includes management leadership, demonstration of employee participation and knowledge of WMSD, hazard identification, job hazard analysis control, and finally employee training. The responsibilities that an organization will accept in the implementation of this process can be classified as

- Management leadership and employee participation
- Job hazard analysis and control
- Training
- WMSD management
- Program evaluation

Each of these areas is discussed and presented in more detail in the following text.

7.12.1   MANAGEMENT LEADERSHIP

Management leadership requires the leadership to take an active and participatory role in addressing and reducing the presence of WMSDs. Specific activities that demonstrate an active concern for employee health and safety include

- Assignment and communication of responsibilities
- Empowering employees to make changes
- Examination of existing policies and practices
- Communication with employees regarding the program
- Prompt response to employee reports of problems

7.12.2 Employee Participation

Employee participation is designed to allow employees to be active participants in the reduction of the risks and occurrence of these injuries. Specific employee participation should include

- Employee participation in the procedure for reporting signs and symptoms of WMSDs
- Access to OSHA standards and information about the ergonomics program

7.12.3 Job Hazard Analysis and Reporting

Job hazard analysis and reporting is designed to identify and remove risks from the workplace. Some of the activities include setting up a way for employees to report WMSD signs and get quick responses as well as the following:

- Provide information to current and new employees:
  - *Common WMSD hazards.*
  - Signs and symptoms of WMSDs.
  - How to report WMSD signs and symptoms.
- Hazard analysis activities will include the following:
  - Evaluation of the problem job to identify "ergonomic risk factors."
  - Implement controls to help eliminate or reduce WMSD hazards.
  - Use an incremental abatement process.
  - When a WMSD hazard occurs, implement one or more controls to materially reduce the WMSD hazards.
  - If WMSD hazard is still occurring, implement additional feasible controls to reduce WMSDs further.
  - Keep implementing controls until the elimination of WMSD hazards.

7.12.4 Training

An important part of a successful ergonomic program is training. The training of employees is essential and should include training employees an follows:

- When the job is identified as a hazard
- When employee is assigned to the problem job
- Periodically, at least every 3 years

Train the persons involved in setting up and managing the ergonomics program

- When initially assigned to setting up and managing the ergonomics program
- Periodically, at least every 3 years

7.12.5 WMSD Management and Program Evaluation

In WMSD management, the employer continually assesses the condition of the workplace and presence of injuries with respect to WMSDs. This includes evaluation of injuries that occur after the implementation of the program and incorporation of additional activities to mitigate risks when necessary. Program evaluation is designed to ensure ongoing effectiveness of the program methods in reducing the risks of injury. To evaluate the ergonomics program an employer must

- Consult with employees in problem jobs to assess their views on the effectiveness of the program
- Identify any significant deficiencies in the program
- Evaluate the elements of the program to ensure they are functioning properly
- Evaluate the program to ensure it is eliminating or materially reducing WMSD hazards

Although this legislation is no longer in effect, it serves as a useful guideline for the establishment of a workplace ergonomics program.

7.13 SUMMARY

Awkward postures, repetitive work, or handling heavy loads are among the risk factors that studies have shown to damage the bones, joints, muscles, tendons, ligaments, nerves, and blood vessels, leading to WMSDs. WMSDs are mostly cumulative, resulting from repeated exposure to loads at work over a period of time. Upper limbs (the hand, wrist, elbow, and shoulder), the neck, and lower back are particularly vulnerable to MSDs. The design of tools, equipment, processes, and work spaces can have a tremendous effect on the risks and occurrence of WMSD.

Despite the growing knowledge of CTDs, according to a survey taken in 1992, the number of CTDs had tripled since 1984. Although a minority of dissenters believes CTDs are a sociopolitical phenomenon with no real scientific grounds, most believe the injuries result from the fact that humans were not built to perform certain tasks requirements at excessive levels. When they do unnatural movements or sit in positions that are not conducive to the human anatomy, the tissue is distorted and swells and eventually begins to tear, causing pain and discomfort. The good news is most CTDs are preventable, if organizations take the time to engineer risk factors out, implement administrative controls and train employees in proper lifting and posturing techniques. Regular job rotation, break enforcement, redesign of the workspace, and personal protective equipment are all preventative measures employers can make to protect their. Companies who manage their CTD risks often find the cost of mitigation will be offset by reduction in medical claims and ultimately a reduction in worker's compensation insurance.

The list in the following text represents ergonomic screening and analysis tools that can be useful in assessing ergonomic risks, particularly for WMSD. These ergonomic tools were obtained from a variety of sources including OSHA; Department of Design and Analysis, Cornell University; Center for Ergonomics at the University of Michigan; Dr. Thomas Bernhard, University of South Florida; and the Human Factors and Ergonomics Society.

TABLE 7.9
Comparison of Five Analysis Tools

Power grip
The hand grip that provides maximum hand power for high force tasks. All the fingers wrap around the handle.

Contact pressure
Pressure from a hard surface, point, or edge on any part of the body.

Pinch grip
The hand grip that provides control for precision and accuracy. The tool is gripped between the thumb and the fingertips.

Single-handle tools
Tube-like tools measured by handle length and diameter. Diameter—the length of a straight line through the center of the handle.

Double-handle tools
Plier-like tools measured by handle length and grip span. Grip span—the distance between the thumb and fingers when the tool jaws are open or closed.

Source: NIOSH website, accessed February 16, 2011.

Checklists and qualitative screening tools

- OSHA computer workstation checklist
- Cornell University performance-oriented ergonomic checklist for computer (VDT) workstation
- Cornell University quick exposure checklist (QEC)
- American Conference of Governmental Industrial Hygienist
  - Threshold limit values for back
  - HAV screening criteria
- WBV screening criteria
- Ergonomic chair evaluation checklist (ANSI/HFES 100-2007)
- Ergonomic work surface evaluation checklist (ANSI/HFES 100-2007)

Ergonomic evaluation tools and techniques

- Physical and postural evaluation tools
  - Manual task risk assessment (ManTRA) tool V2.0
  - Rapid upper limb assessment (RULA) tool
  - Rapid entire body assessment (REBA) tool
  - JACK—human simulation and ergonomics analysis software
  - University of Michigan, 3D Static Strength Prediction Program (3DSSPP)
  - Job strain index (JSI)
  - Moore–Garg strain index
  - Rohmert method for static work analysis
- Cornell University ergonomic tools and techniques
  - Ergonomic seating evaluation form
  - Ergonomic keyboard and mouse platform systems evaluation form
  - Ergonomics cost-justification worksheets
  - Cornell return on investment (ROI) estimator
  - Cornell University Body Discomfort Survey 2010
  - Cornell musculoskeletal discomfort questionnaires

Case Study

Study looks at potential effects of multi-touch devices
By
Kanav Kahol, Ph.D.
June 9, 2010

Kanav Kahol, an assistant professor in ASU's Biomedical Informatics Department, is leading team of researchers in a project to measure the amount of strain on the hands and wrists of individuals who use multi-touch electronic

devices, like Apple's iPad. They will use cyber gloves to measure the kinematic features produced while interacting with multi-touch systems. ASU photo by Scott Stuk.

The evolution of computer systems has freed us from keyboards and now is focusing on multi-touch systems, those finger-flicking, intuitive and easy-to-learn computer manipulations that speed the use of any electronic device from cell phones to iPads. But little is known about the long-term stresses on our bodies through the use of these systems.

Now, a team of researchers led by Kanav Kahol of ASU is engaged in a project to determine the effects of long-term musculoskeletal stresses multi-touch devices place on us. The team, which includes computer interaction researchers, kinesiologists and ergonomic experts from ASU and Harvard University, also are developing a tool kit that could be used by designers when they refine new multi-touch systems.

"When we use our iPhone or iPad, we don't naturally think that it might lead to a musculoskeletal disorder," said Kahol, an assistant professor in ASU's Department of Biomedical Informatics. "But the fact is it could, and we don't even know it. We are all part of a large experiment. Multi-touch systems might be great for usability of a device, but we just don't know what it does to our musculoskeletal system."

As we move towards a world where human-computer interaction is based on various body movements that are not well documented or studied we face serious and grave risk of creating technology and systems that may lead to musculoskeletal disorders (MSD), Kahol said.

Many of today's multi-touch systems have no consideration of eliminating gestures that are known to lead to MSDs, or eliminating gestures that are symptomatic of a patient population, Kahol said. This project—supported by a $1.2 million grant from the National Science Foundation—aims to develop best practices and standards for interactions that are safe and cause minimal user stress while allowing users to fully benefit from the new levels of immersion that multi-touch interaction facilitates.

In addition to Kahol, co-principal investigators on the project are Jack Dennerlein of the Harvard School of Public Health, Boston, and Devin Jindrich, an ASU kinesiologist.

Kahol said the project initially will focus on evaluating the impact multi-touch devices have on the human musculoskeletal system. Users will be fitted with electromyography (EMG) equipment to measure muscle forces, and cyber gloves to measure kinematic features that are produced while they interact with multi-touch systems. The researchers will then evaluate the impact of those stresses.

The second part of the project will develop biomechanical models where the user will be able to "enter the motion of a gesture, and the system will produce the forces being exerted through that motion, like a specific movement of the hand," Kahol explained. "We would then take this data back to the Microsofts, the Apples and other manufacturers so they could use it when they are designing new devices."

The system, Kahol said, will be built with off the shelf components and it will give designers a new tool to use when developing new multi-touch systems.

"The designers, the computer scientists, the programmers, they know little about biomechanical systems, they just want a system that they can employ in a usable manner and tells them if a gesture causes stress or not," Kahol said. "So our major challenge is going to be developing the software, the tool kit and the underlying models that will drive the tool kits." Kahol said that the last time designers developed a fundamental interaction system with computers they modified the standard keyboard. While it was useful, it was not without its share of drawbacks.

"When we developed the keyboard, we didn't think through how working with it would affect the hands, arms, etc.," Kahol said. "As a result, it created a multimillion dollar industry in treating carpal tunnel syndrome. That is what we want to prevent with multi-touch systems. "We are going for the preventative, rather than the curative," he added.

Source: Total contents of Case Study reprinted with permission from Ergoweb.com

EXERCISES

7.1 Discuss the history of work-related musculoskeletal injuries in your country and on a global level.

7.2 Through which body parts is vibration most likely transmitted?

7.3 Is it reasonable to assume that vibration is transmitted to the body either only horizontally or only vertically, as is done in most research?

7.4 What professionals other than truck drivers are particularly exposed to vibrations or impacts?

7.5 What might be the effects of several sources of vibration arriving simultaneously at different body parts?

7.6 Which of the following tasks might provoke symptoms in a worker suffering from carpal tunnel syndrome? List expected risk factors associated with each task.

    a. Scrubbing debris from a brick wall using a wire brush held in the hand

    b. Operating a computer keyboard

    c. Painting a ceiling

    d. Mopping a floor

    e. Playing tennis

7.7 Which of the following jobs might be "strongly" associated with anterior knee pain later in life?

    a. Sailor

    b. Tiler

    c. Chef

    d. University lecturer

    e. Administrator

    f. Nurse

    g. Farmer

    h. Doctor

7.8 The occupational medical officer in a poultry processing plant wishes to reduce the high incidence of WMSDs. Which of the following would be best to focus on first?

    a. Aerobic capacity of the workers

    b. Lighting levels

    c. Prevalence of repetitive tasks

    d. Job contentment

7.9 Which of the following alterations to the design of a powered tool would be most likely to reduce the prevalence of WMSDs?

    a. Fitting a high-friction sleeve to the handle

    b. Changing from a pinch grip to a power grip

    c. Switching from foot pedal operation to a hand-operated trigger

    d. Designing a control that requires continuous force to operate

    e. Increasing the power of the tool

7.10 List two occupational and two nonoccupational risk factors for CTS.

7.11 Which of the following might be caused by working all day at a computer with the mouse positioned too far to one side of the user and at an elevated position?

    a. Pain in the lumbar region of the back

    b. Neck pain

    c. Pain on the opposite side of the body

d. Rotator cuff tendonitis
e. CTS
f. Wrist tendonitis
g. RSI
h. Nonspecific forearm pain

REFERENCES

Armstrong, T. (2002). *Upper Limb Musculoskeletal Disorders and Repetition*, University of Michigan, Ann Arbor, MI, http://www-personal.umich.edu/~tja/JobEval/analysis2.html

ANSI (1996). American national standard: Control of work-related cumulative trauma disorders. Part I: Upper extremities. American National Standard Institute, New York, 2–365.

Bernard, B. (ed.) (1997). *Musculoskeletal Disorders and Workplace Factors: A Critical Review of Epidemiologic Evidence for Work-Related Musculoskeletal Disorders of the Neck, Upper Extremity, and Low Back*, Publication No. 97B141, DHHS (NIOSH): Cincinnati, OH.

Bernard, B., Sauter, S., Fine, L.J., Petersen, M., and Hales, T. (1994). Job task and psychosocial risk factors for work-related musculoskeletal disorders among newspaper employees. *Scand. J. Work Environ. Health*, 20(6), 417–426.

Buckwalter, J.A., Woo, S.L.-Y., Goldberg, V.M., Hadley, E.C., Booth, F., Oegema, T.R. et al. (1993). Current concepts review: Soft-tissue aging and musculoskeletal function. *J. Bone Joint Surg. Am.*, 75A(10), 1533–1548.

Bureau of Labor Statistics. (2004). Tables R9–R12: Occupation × selected case characteristics. http://www.bls.gov/iif/oshcdnew.html "04b" www.bls.gov/iif/oshcdnew.htm#04b (accessed: September 2011).

Bureau of Labor Statistics. (2008). Table R8. Incidence rates for nonfatal occupational injuries and illnesses involving days away from work per 10,000 full-time workers by industry and selected events or exposures leading to injury or illness, 2007. Bureau of Labor Statistics, U.S. Department of Labor: Washington, DC, Survey of Occupational Injuries and Illnesses in cooperation with participating State agencies.

Bystrom, S., Hall, C., Welander, T., Kilbom, A. (1995). Clinical disorders and pressure-pain threshold of the forearm and hand among automobile assembly line workers. *J. Hand Surg. Br.*, 20B(6), 782–790.

Canadian Occupational Safety and Health, http://www.ccohs.ca/oshanswers/diseases/rmirsi.html

Chiang, H.-C., Yin-Ching, K., Chen, S.-S., Hsin-Su, Y., Trong-Neng, W., Chang, P.-Y. (1993). Prevalence of shoulder and upper-limb disorders among workers in the fish-processing industry. *Scand. J. Work Environ. Health*, 19, 126–131.

Collins, J., Nelson, A., and Sublet, V. (February 2006). *Safe Lifting and Movement of Nursing Home Residents*, Publication No. 2006-117, DHHS (NIOSH): Cincinnati, OH.

David, W. and Chaudhry, V. (2003, March). Literature review: Nervepace digital electroneurometer in the diagnosis of carpal Tunnel syndrome. *Muscle Nerve*, 27(3), 378–385.

Durkan, J.A. (1991, April). A new diagnostic test for Carpal Tunnel syndrome. *J. Bone Joint Surg.*, 73(4), 535–537.

European Agency for Safety and Health at Work. (2000). Research on work-related stress, http://osha europa eu/publications/reports/203/stress_en pdf

European Agency for Safety and Health at Work, Facts 22: Work-related stress, http://osha europa eu/publications/factsheets/22/factsheetsn22_en pdf

European Agency for Safety and Health at Work. (2009). Outlook 1—New and emerging risks in occupational safety and health, Office for Official Publications of the European Communities: Luxembourg, http://osha.europa.eu/en/publications/outlook/new-and-emerging-risks-in-occupational-safety-and-health-annexes (accessed: September 2011).

European Agency for Safety and Health at Work. (2003). Improving occupational safety and health in SMEs: Examples of effective assistance, http://osha.europa.eu/en/publications/factsheets/37

Hales, T.R., Sauter, S.L., Peterson, M.R., Fine, L.J., Putz-Anderson, V., Schleifer, L.R. et al. (1994). Musculoskeletal disorders among visual display terminal users in a telecommunications company. *Ergonomics*, 37(10), 1603–1621.

Holmstrom, E.B., Lindell, J., Moritz, U. (1991). Low back and neck/shoulder pain in construction workers. Occupational workload and psychosocial risk factors. *Spine*, 17, 663–677.

Holmström, E.B., Lindell, J., Moritz, U. (1992). Low back and neck/shoulder pain in construction workers: Occupational workload and psychosocial risk factors. Part 2: Relationship to neck and shoulder pain. *Spine*, 17(6), 672–677.

Jetzer, T.C. (1991, February). Use of vibration testing in the early evaluation of workers with Carpal Tunnel syndrome. *J. Occup. Med.*, 33(2), 117–120.

Johansson, J.A., Rubenowitz, S. (1994). Risk indicators in the psychosocial and physical work environment for work-related neck, shoulder and low back symptoms: A study among blue and white collar workers in eight companies. *Scand. J. Rehabil. Med.*, 26, 131–142.

Karwowski, W. and Marras, W.S. (eds.) (1998). *The Occupational Ergonomics Handbook*, CRC Press: Boca Raton, FL.

Kroemer, K.H.E., Kroemer, H., and Kroemer-Elbert, K. (1994). *Ergonomics: How to Design for Ease and Efficiency*, Prentice Hall: Englewood Cliffs, NJ.

Kuorinka, I. and Forcier, L. (Eds.) (1995). *Work Related Musculoskeletal Disorders (WMSDs): A Reference Book for Prevention*, Taylor & Francis: Boca Raton, FL.

Letter by Charles N. Jeffress, April 29, 1999; regarding OSHA Ergonomics legislation; OSHA website, http://www.osha.gov/pls/oshaweb/owadisp.show_document?p_id=236&p_table=SPEECHES (accessed: September 2011).

Mandel, M.A. (2011). Cumulative trauma disorder history, pathogenesis and treatment, http://www.workinjuryhelp.com/ouch/mandel.htm (accessed: February 2011).

Marras, W.S., Lavender, S.A., Leurgans, S.E., Fathallah, F.A., Ferguson, S.A., Allread, W.G. et al. (1995). Biomechanical risk factors for occupationally-related low back disorders. *Ergonomics*, 38(2), 377–410.

McCauley-Bell, P. and Badiru, A. (May 1996). Fuzzy modeling and analytic hierarchy processing to quantify risk levels associated with occupational injuries—Part I: The development of fuzzy linguistic risk levels. *IEEE Trans. Fuzzy Syst.*, 4, 124–131.

National Archives. (2011). Teaching with documents: Photographs of Lewis Hine: Documentation of child labor, http://www.archives.gov/education/lessons/hine-photos

National Safety Council (NSC). (1990). *Accident Facts*, NSC: Chicago, IL.

Neese, R. and Konz, S. (1993, July). Vibrometry of industrial workers: A case study. *Int. J. Ind. Ergon.*, 11, 341–345.

NIOSH. (1997). *Musculoskeletal Disorders and Workplace Factors*, NIOSH Publication No. 97-141, http://www.cdc.gov/niosh/docs/97-141

OSHA Website, Guidelines for Nursing Homes Ergonomics for the Prevention of Musculoskeletal Disorders. (2009). U.S. Department of Labor, Occupational Safety and Health Administration, Publication # OSHA 3182-3R 2009 (accessed February 18, 2011).

Pransky, G., Long, R., Hammer, K., Schulz, L., Himmelstein, J., and Fowke, J. (1997, August). Screening for Carpal Tunnel syndrome in the workplace: An analysis of portable nerve conduction devices. *J. Occup. Environ. Med.*, 39(8), 727–733.

Ranney, D., Wells, R. and Moore, A. (1995). *Ergonomics*, 38(7), 1408.

Sauter, S., Hales, T., Bernard, B., Fine, L., Petersen, M., Putz-Anderson, V., Schleiffer, L., and Ochs, T. (1993). Summary of two NIOSH field studies of musculoskeletal disorders and VDT work among telecommunications and newspaper workers. In: H. Luczak, A. Cakir, and G. Cakir (eds.), Elsevier Science Publishers, B.V.

Silverstein, B.A., Fine, L.J., and Armstrong, T.J. (1987). Occupational factors and the carpal tunnel syndrome. *Am. J. Ind. Med.*, 11(3), 343–357.

The Burden of Musculoskeletal Disease in the United States. (2011). http://www.boneand jointburden.org (accessed: September 2011).

Urbano, F.L. (2000, July). Tinel's sign and Phalen's maneuver: Physical signs of Carpal Tunnel syndrome. *Hosp. Physician*, 39–44.

U.S. Department of Health and Human Services, Office on Women's Health, http://www. womenshealth.gov/faq/carpal-tunnel-syndrome.cfm (accessed: April 2011).

Werner, R.A., Franzblau, A., and Johnston, E. (1994, November). Quantitative vibrometry and electrophysiological assessment in screening for Carpal Tunnel syndrome among industrial workers: A comparison. *Arch. Phys. Med. Rehabil.*, 75(11), 1228–1232.

White, K., Congleton, J., Huchingson, R., Koppa, R., and Pendleton, O. (1994, January). Vibrometry testing for Carpal Tunnel syndrome: A longitudinal study of daily variations. *Arch. Phys. Med. Rehabil.*, 75(1), 25–27.

Women in America: Indic   ators of Social and Economic Well-Being. (2011). Prepared by U.S. Department of Commerce Economics and Statistics Administration and Executive Office of the President, Office of Management and Budget.

8 Heavy Work and Evaluating Physical Workloads and Lifting

8.1 LEARNING GOALS

Understanding the effects and problems associated with manual material handling (MMH) of heavy loads is critical to task design in such environments. Factors such as physical strength, fatigue, temperature, slopes, and load all affect the efficiency and safety of the tasks. Long-term health risks, primarily musculoskeletal disorders, are common performance in such task. Mitigating risks associated with these types of physically intensive tasks includes understanding how to use evaluation tools such as the National Institute for Occupational Safety and Health (NIOSH) lifting guideline worker limitations and effective task design.

8.2 KEY TOPICS

- Heavy work
- MMH and lifting
- Assessment of energy expenditures: direct and indirect measures
- Evaluation of physical efforts in MMH
- NIOSH lifting guideline
- Fatigue

8.3 INTRODUCTION AND BACKGROUND

In George Orwell's *Down and Out in Paris and London*, a largely autobiographical tale, the author details the life of a young laborer taking work as a dishwasher, waiter, and other odd restaurant and hotel jobs, the life of a common worker in the early 1900s. Working up to 18 h a day, Orwell describes how "he lives in a rhythm between work and sleep, without time to think, hardly conscious of the exterior world [...] Nothing is quite real to him but the *boulot*, drinks and sleep; and of these sleep is the most important" (Orwell, 1961).

Although much has changed since this novel was originally published in 1933 and over the last 30 years special attention has been given to the necessity of sleep for the health and efficiency of the workforce. Humans (and most other animals for that matter) have a natural tendency to sleep during nighttime hours and are more energetic during the day. This is related to our circadian rhythm, an internal clock that is guided primarily by, among other factors, natural daylight—the sun. One of the

most notorious examples of improper attention to circadian rhythm and sleep awareness was the Exxon Valdez oil spill in March 1989. This accident, the largest spill
in U.S. history at the time, occurred in Alaska's Prince William Sound, well north
of the point where the sun never sets. For this reason, the shipmates, who, according to one report, were already battling "environmental factors such as long work
hours, poor work conditions (such as toxic fumes), monotony and sleep deprivation"
were probably unable to adjust their circadian rhythm to the longer daylight hours in
that location, should they have had the opportunity to sleep as long as they wanted
(Exxon Valdez Oil Spill Trustee Council, 1990).

This issue also becomes a problem for shift workers, who work at night and have
to adjust their sleep schedule to accommodate their work schedule. Recent studies
have shown that the best kind of light fixtures for keeping nightshift workers alert is
blue light, or shortwave light, which appears blue. Another way in which businesses
can help their employees adjust their sleep schedules is by introducing a flextime
work schedule. Under a flextime policy, employees will work a determined number
of hours within a fixed period of time (e.g., between 9 a.m. and 1 p.m.) but are able to
vary the time they start or finish work (Washington State Legislature, 1985).

In addition to fatigue due to unnatural sleeping habits, heavy industry, like construction, steel manufacturing, shipbuilding, and baggage handling, adds new dangers to the equation. Therefore, it is especially important that managers provide tools
to prevent injury, like back belts and coupling devices, as well as offer training on
proper lifting techniques. To determine the maximum load a worker can lift, the
NIOSH lifting equation is used. This tool takes into account "the weight of the object
being lifted, the horizontal and vertical hand locations at key points in the lifting
task, the frequency rate of the lift, the duration of the lift, the type of handhold on
the object being lifted, and any angle of twisting" (ErgoWeb).

8.4   HEAVY WORK

Although physically intensive task remain in some industries, heavy work is becoming increasingly rare due to mechanization industry focus toward more information-
oriented tasks, and recognition of the risks of strenuous tasks. Generally, heavy work
is characterized by extensive muscular exertion in highly physical task activities.
Industries with heavy physical work include: firefighters, carpentry, construction
workers, farm workers, professional football and basketball players, and many more.
Intensive work may place stress on all of the systems of the body including

- *Muscular*—Extensive force exertion, load balancing, and physical activity
- *Cardiovascular*—Need to pump blood to muscles as work is performed
- *Respiratory*—Need to perform gas exchange to carry nutrients to muscles
  and remove waste products
- *Nervous system*—Through needed innervations of muscles for required
  movements and motor control

A number of ergonomic concerns must be addressed when effectively designing
tasks that involve heavy work.

8.4.1 Energy Consumption during Heavy Work

During the past several years, automation of many tasks has greatly reduced the demands for strength and energy needed to perform certain tasks. Yet, there are still many industries with jobs that require very heavy work from their workers. Heavy work is characterized by any activity that requires substantial physical exertion and results in high-energy expenditure and extensive pressure on the heart and lungs. The consumption of energy is measured in kilojoules (kJ). The severity of a physical task is often measured by the consumption of energy and cardiac effort exerted during the task. To effectively assess task demands for physically intensive jobs, dynamic task analysis is necessary. During static efforts, blood flow is insufficient and the contracted muscles are supplied by comparatively little energy. Although static work can be exhausting, very little energy consumption takes place at the contracted muscles. Therefore, it is difficult to assess energy measures given the static and short nature of these exertions.

People who are well capable and trained can perform light, medium, and even heavy work with their metabolic and other physiological functions reaching a stable safe level throughout the physical work period. However, this is not the case with very heavy work. During very heavy work tasks, energy expenditures can reach 40 kJ/min and the heart rate (HR) is approximately 140 beats/min (Kroemer and Grandjean, 1997). Furthermore, oxygen deficit increases for the entire duration of the work, either requiring the person to take several rest periods or even forcing them to completely stop working. Higher energy expenditures of 50 kJ/min and HRs of 160 beats/min or more cause an enormous increase of lactic acid in the blood, as well as drastic increases of oxygen deficit. During such extremely heavy work performances, frequent rest periods are inevitable and even well-trained and capable persons may not be able to accomplish the job through a full work shift. During heavy physical work, several adjustments and adaptations take place within the body that affect almost all the organs, tissues, and fluids of the body. Some of the most important adjustments are summarized in the following:

- Deeper and more rapid breathing.
- Increased HR, accompanied by an initial rise in cardiac output.
- Vasomotor adaptations, with dilation of the blood vessels in the organs involved, such as the muscle and heart, while other blood vessels are constricted. This diverts blood from the organs not immediately required into those that need more oxygen and nutrients.
- Rise in blood pressure, increasing the pressure gradients from the main arteries into the dilated vessels of the working organs and so speeding up the flow of blood.
- Increased supply of sugars including glucose and glycogen released into the blood by the liver.
- Rise in body temperature and increased metabolism. The increase in temperature speeds up the chemical reactions of metabolism and ensures that more chemical energy is converted into mechanical energy.

8.4.2 Energy Efficiency of Heavy Work

Even though heavy work is becoming less common in industrial countries, it is still prevalent in some occupations, particularly in developing countries. Therefore, it is very important in these countries to apply ergonomics with a high level of efficiency in heavy work. Human physical effort is not very efficient with only 30% of effort going toward a given physical exertion. This implies that 30% of the produced energy during metabolism is consumed into mechanical work (or the task being performed) and the remaining 70% of the energy is converted into heat. The highest level of efficiency can be achieved by converting as much of the mechanical energy as possible into useful work without allowing much of this energy to go into holding or supporting objects.

Proper handling of heavy work also involves finding ways to minimize any type of stress that the worker endures while executing a task and designing the task to promote efficiency. For a task to be designed to be efficient, the psychological, physical, and environmental factors must be considered. The categories and basic factors within each are listed in the following text:

1. Psychological factors
   a. Task difficulty
   b. Psychological stress and/or time pressure
   c. Cognitive load: decision making and perceptual work
2. Physical factors
   a. Task duration
   b. Work posture
   c. Physical effort (force and torque)
   d. Manual material handling
      i. Lifting
      ii. Carrying
      iii. Pushing
3. Environmental factors
   a. Heat, cold, humidity
   b. Chemicals, noise

The metabolic demands of industrial task are used to classify the level of physical demands of the task. The classification of work ranges from light to extremely heavy (Eastman Kodak Company, 1986) and examples of these levels of work are shown in Table 8.1.

Several industrial jobs and tasks are classified according to their average energy demands, ranging from light (column 1) to extremely heavy (column 5) work. The range of energy expenditures is shown at the top of each column, in watts and kilocalories per hour. Tasks in the light- and moderate-effort categories (columns 1 and 2) are more likely to be done for a full 8 h shift. Heavy to extremely heavy tasks (columns 3–5) are usually alternated with more sedentary paperwork or standby activities.

TABLE 8.1
Metabolic Demands of Industrial Tasks[a]

Light	Moderate	Heavy	Very Heavy	Extremely Heavy
70–140 W (60–120 kcal/hr)	>140–280 W (>120–240 kcal/hr)	>280–350 W (>240–300 kcal/hr)	>350–420 W (>300–360 kcal/hr)	>420 W (>360 kcal/hr)
Small-parts assembly	Industrial sewing	Making cement	Shoveling (>7 kg)	Stoking a furnace
Typing	Bench work	Industrial cleaning	Ditch digging	Ladder or stair climbing
Keypunching	Filing	Joining floorboards	Hewing and loading coal	Coal car unloading
Inspecting	Machine tending	Plastering	Handling moderately heavy (>7 kg) cases to and from a pallet	Lifting 20 kg cases 10 times per minute
Operating a milling machine	Small-size packing	Power truck operation	Tree planting	
Drafting	Operating a lathe	Handling light cases to and from a pallet	Handling heavy (>11 kg) units frequently (>4 per min)	
Armature winding	Operating medium size presses	Road paving		
Hand typesetting	Machining	Painting		
Operating a drill press	Small-sized sheet metal work	Handling operations		
Desk work	Electronics testing	Metal casting		
Small-parts finishing	Plastic molding	Cutting or stacking lumber		
	Operating a punch press	Large-size packing		
	Operating a crane laying stones and bricks			
	Sorting scrap			

* The values given include basal metabolism.

8.4.3 EFFECTS OF HEAVY WORK AND HEAT

Extreme heat contributes to the impact of heavy work and changes in HR can be used to predict this impact. HR is widely used in studies as a measure of physiological demand and workload because the heart, which acts as a pump to supply the muscles with blood as well as supporting cooling and respiration through the skin and lungs.

The primary functions of the heart while performing heavy work under extremely hot conditions include

- Transporting energy to the muscles
- Transporting heat energy from the interior of the body to the skin to cool the body
- Supporting increasing needs in respiration

Examples of industries where workers endure these types of conditions are varied and include, but is not limited to, construction, agriculture, and maritime industries, Occupational Safety and Health Administration (OSHA) has guidelines and resources available to support the analysis of heat load and reduce the risk of overheating in the occupational environments.

8.4.4 GROUND SURFACES, SLOPES, AND STAIRCASES

The type of ground surface where work is being performed has a definite impact on energy consumption. For example, walking on smooth and solid ground is not as strenuous as walking on dirt road, light or heavy brush, hard-packed snow or swampy bog. Making use of a wheelbarrow or other type of cart for moving loads when working under these conditions can be extremely beneficial. It is more efficient to push a cart since the effort requirement is 15% less compared to dragging it from behind. It is recommended that the handles of the cart be about 1 m above the ground and approximately 40 mm thick. Also, the center of mass of a two-wheeled cart should be as low and as close to the axle as possible for stability and to minimize the amount of weight that is supported.

Jobs that required climbing should be performed at a slope of approximately 10° since it generates the best efficiency. Climbing ladders are most efficient when placed at an angle of 70° and the rungs are 260 mm apart. However, if handling heavy loads, a ladder with 170 mm between the rungs is preferred. One of the best forms of exercise involves climbing stairs. An efficient design of staircases is very important, particularly when the stairs are frequently used or used by older people. Lehmann suggested that staircases with a slope between 25° and 30° consumed the least energy (Lehmann, 1962). Furthermore, he proposed two empirical norms when designing staircases since these dimensions proved to be most efficient and cause fewer accidents.

- The tread height or riser should be 170 mm.
- The tread depth should be 290 mm.

Additional guidelines are available on the OSHA website.

8.4.5 EVALUATION OF PHYSICAL WORKLOADS

A number of approaches exist for assessment of physical workload or manual exertions. These techniques include mathematical models, databases, analysis or prediction software, and physiological evaluation tools. Table 8.1 can be used as a guideline to determine the type of technique that is most appropriate for manually intensive tasks.

The two broad categories for these approaches include physiological and psychophysical. A description of ergonomic approaches to perform the evaluation of physical and psychological evaluation techniques.

- Physiological
  - Identifies limitations in central capabilities (pulmonary, circulatory, or metabolic functions)
  - Provides essential criteria if functions are highly taxed my MMH job
- Psychophysical
  - Addressed all (local and central) functions strained in the test
  - May include all or many of the systems evaluated in the other examinations
  - Filters the strain experienced through the sensation of the subject who rates the perceived exertion
  - Subject decides how much strain is acceptable under given conditions
- Physiological instrumentation: HR
  - Heart beat frequency (heart rate) is the most important and simplest parameter of cardiovascular system function
  - Physical work evokes acceleration of HR and an increase in the volume of blood ejected by the heart per unit of time (cardiac output)
  - Changes in cardiovascular system function allow the delivery of more oxygen and nutrients to active tissues
- Physiological instrumentation: energy expenditure
- Energy expenditure can be measured directly (calorimetric chambers) or indirectly (oxygen consumption during exercise)
- Oxygen consumption measured with electronic oximeters
  - VO_2Max is the most amount of oxygen that can be used by your body (muscles) in 1 min. It is measured in units of milliliters per kilogram of body weight per minute
- Energy expenditure is calculated on the basis that oxygen consumption is the result of multiplying oxygen consumption (in liters per minute) by the oxygen energy equivalent, taking into account the respiratory quotient (RQ) (ratio of carbon dioxide to oxygen in expired air)
  - The result is expressed in kJ/min

$$\text{Energy expenditure (kJ/min)} = VO_2(1/\text{min})$$
$$\times \text{energy equivalent of } O_2 \text{ (for a given RQ)} \qquad (8.1)$$

- Physiological load and work tolerance depend on the individual working capacity
  - The suggested physical workload is 30% VO_{2max} during a shift
  - Load that exceeds 50% VO_{2max} per shift can adversely affect the workers' health in the long run

Shoveling, which is very common in manual labor within many industries, has been studied for many years due to the strenuous nature of the activity. Additionally, shoveling of snow in cold climates is a risk factor that many individuals experience at home. Studies have revealed that certain forms of shoveling tools produce more efficient work with less effort than others. When performing tasks that involve shoveling, it is important to consider the following (Coffiner, 1997):

- Consider the weight of the shovel in the analysis of the load.
- A big shovel is recommended for light materials and a small shovel should be used for heavy materials.
- Fine-grained materials should be shoveled with a slightly hollow, spoon-shaped with a pointed tip shovel, while coarse materials require a shovel that has a straight cutting edge and a flat blade, with a rim around the back and sides.
- A shovel with a cutting edge that is either straight or pointed but with a flat blade is highly recommended for stiff materials, such as clay.
- The length of the handles of both spades and shovels should be between 600 and 650 mm.

Another tool that is widely used in manual work is the saw. Studies of several types of saws have shown that the preferred saws are ripsaws, which a have a broad blade and either two or four cutting teeth in each group. Other studies suggested that efficient results when using a saw were obtained from a sawing rate of 42 double strokes/min and a vertical force of about 100 N. When hoeing in soft soil, a swivel hoe is preferred over the ordinary chopping hoe. However, if the soil is both hard and dry, the two types of hoes are equally good.

Efficient walking speeds have been studied for several years. A walking pace that is comfortable and not too strenuous is between 75 and 110 steps/min, with a length of pace between 0.5 and 0.75 m. Yet, this is not the most efficient pace when the work performed and energy consumed are compared against each other. Studies by Hettinger and Muller (HeHinger and Muller, 1953) revealed that the most efficient walking speed is between 4 and 5 km/h, but this speed is reduced to 3 and 4 km/h when wearing heavy shoes. Special attention has been given to more strenuous forms of work, such as those that involve carrying heavy loads. Although Lehmann found the most efficient load between 50 and 60 kg, smaller loads are highly recommended when possible, as they are more convenient and safe. Carrying smaller loads does require more trips to and from, and may result in an increase in the consumption of energy. Another option is to distribute the carried weight into smaller proportions: and allow it be carried differently for example, to the back, the shoulders, the chest, or even on the head. The least energy-consuming load is about 30 kg, with the weight distributed on the back and chest. This explains why it's easier to carry a comparable load as a back pack vs, on the shoulder.

8.4.6 VO₂MAX

VO_2Max is a measure of the body's capacity for aerobic work and, thus, can be a predictor of task performance endurance athletes or in physically intensive tasks such

TABLE 8.2

Oxygen Uptake and Physical Activity

Rest	Basal metabolism require approximately 0.25 L of oxygen per minute
Sedentary work	Office work, such as oxygen uptake of only a little over resting levels (0.3–0.4 L/min)
Housework	Housework includes several moderate to heavy tasks (requiring about 1 L/min of oxygen). It is unusual for high work rates to be maintained for any length of time
Light industry	Oxygen uptakes from 0.4 to 1 L/min are required
Manual labor	Oxygen uptake may vary from 1 to 4 L/min. The workload can depend greatly on the tools and methods (heavy MMH)
Sports	Endurance sports or very occupational tasks that take over 5 L/min

Source: Bridger, 1995.

as firefighting. Technically, VO_2Max stands for maximal oxygen uptake and refers to the amount of oxygen your body is capable of utilizing in 1 min. The units are: ml O_2/kg min—meaning milliliters of oxygen per kilogram of body weight per minute. Other factors that affect aerobic capacity include training, genetics, body weight, muscle volume, etc. Age is also a factor, as most people see a decline of about 1% a year in VO_2Max after age 50.

Accurate measurement of VO_2Max generally requires the collection and analysis of inhaled and exhaled gases during exercise to exhaustion. This is usually done in a lab, under the supervision of professionals. Due to the complexity of obtaining VO_2Max values, these values are often approximated in occupational environments based on employee age and general physical condition. Likewise, measured VO_2Max levels are often used to classify aerobic fitness levels. Table 8.2 provides the ranges for classifying aerobic fitness levels using VO_2Max.

8.5 MANUAL MATERIAL HANDLING AND LIFTING (MMH)

MMH, including lifting tasks, contributes to a large percentage of the over-half-a-million cases of musculoskeletal disorders reported annually in the United States (OSHA, 2011). Risk factors include work related musculoskeletal disorders (WMCD), occlusion of vision (if the load is large and blocks the view), as well as a likelihood for trips, slips, and falls. Load handling tasks are in numerous industries including manufacturing, construction, health care, and others. The variability in these occupational environments means that any guidelines for managing MMH should be broad and allow for adaptation to the specific needs of the given workers. OSHA and NIOSH offer guidelines, research, and publications that can be useful across different industries as employers seek to minimize the impact and risks of these MMH on the worker.

8.5.1 MATERIAL HANDLING

Handling material involves any type of manipulation of an object or load. The object or load may be soft or solid, bulky or small, smooth or with corners and edges, and

FIGURE 8.1 Manual Material Handling (Source: Ergonomic Guidelines for Manual Material Handling: http://www.cdc.gov/niosh/docs/2007-131/pdfs/2007-131.pdf)

may also include handling humans or pets. Figure 8.1 is a common example of manual material handling. Handling material requires exertion of energy or force for pushing, pulling, lifting, lowering, dragging, carrying, and holding objects. The three types of material handling that are mainly studied include:

1. Lifting, which means moving an object by hand from a lower position to a higher one
2. Lowering, which is just the opposite of lifting
3. Pushing, pulling, carrying, and holding an object

The energy involved in the handling of loads must be produced within the body and released in terms of force or torque. Material handling tasks are the primary cause of back-related injuries in occupational settings. In fact, low back pain is recognized as the leading cause of morbidity and lost productivity in the workforce today. Some estimates say that as many as 60%–70% of people in physically intensive tasks have suffered from back problems. Such injuries can be found in sports, especially in weight lifting, as well as, during leisure and such as gardening activities. The widespread occurrence of low back injuries has been a main concern for a long time, urging ergonomic intervention to help prevent or at least reduce the risk of overexertion in people handling materials.

8.5.2 Classification of Manual Material Handling Characteristics

Understanding the factors that are associated with MMH characteristics tasks can help minimize the risks that these tasks involve. A systems approach should be applied to the design of material handling tasks to holistically understand the task environment. The four categories that are used to classify MMH task characteristics are (Mital et al., 1993) as follows:

- Task characteristics
- Material/container characteristics
- Work practices characteristics
- Worker characteristics

8.5.2.1 Task Characteristics

The task requirements should be evaluated to thoroughly understand the material handling expectations associated with a task. Confined or limited workspaces can be a risk factor for low back pain. Work in spaces that constrain an individual's posture should be eliminated where possible, especially when it affects headroom or horizontal reaches. If the task requires the operator to reach over obstacles and into containers at a distance from the torso, this can produce stress and strain on the upper extremities and back. Thus the work area should be unobstructed to allow the worker ample space for movement in picking up, handling, and placing of an object. Foot and legroom should be sufficient to allow the worker movement for bending the legs and knees when getting close to the object for picking up or placement. Finally, floors should be free of debris or materials that might pose a slip, trip, or fall hazard. Many materials handling jobs are performed while standing. It is generally agreed that there is a correlation between standing for extended periods of time (4 h or more) and low back pain. Generally speaking, the harder the flooring (with concrete being the worst for the lower back), the more discomfort and fatigue are likely. Flooring properties, surface treatments, and shoe sole materials need to provide ample friction between the shoes and floor, especially when the job requires heavy lifting or when materials are to be pushed or pulled. Adequate floor/foot friction should be provided to improve effort efficiencies and reduce the chance of slipping (NC OSHA, 2009).

To summarize, the following factors should be evaluated in assessing task characteristics:

- Workplace geometry, spatial properties of the task during picking up, handling, and placing of load
- Frequency, duration, and pace of task
- Complexity or precision or load handling and placement
- Environmental factors including noise, temperature, vibration, and other conditions
- Type of flooring and impact on worker during load handling

8.5.2.2 Material/Container Characteristics

Generally speaking, when it comes to the manual handling of loads, smaller and lighter loads are easier to handle. Large, awkward loads present the worker with a variety of potential problems including added stress and strain to the upper extremities and the back. Containers should not be so tall that vision is obstructed during lift or carry or conversely so long that the container makes contact with the legs as it is are carried. Loads should be designed so they can be carried close to the spine, thereby reducing the load's compressive forces on the lower back.

Likewise, loads should not be too light. Loads that are too light may encourage the worker to lift multiple units at a time, creating an unstable load that is more likely to fall. Containers should be designed with stable loads to prevent their contents from shifting. Loads that shift in their containers can move the center of gravity away from the handler, suddenly creating a risk of dropping the load and increasing the load on the lower back. The edges of the container should be rounded, not sharp, to eliminate risk of contract stress between the load and the hand, arm, and body. Whenever possible, containers should have good coupling. Handles or hand cutouts provide the best coupling between the handler and the object. According to the 1991 Revised NIOSH Lifting Equation (NC OSHA, 2009, design parameters have been converted from inches to centimeters)

- Handles should be of a cylindrical shape with a smooth, nonslip surface
- Handle design
  - 1.9–3.8 cm in diameter
  - A minimum of 11.5 cm long
  - 5.0 hand clearance
- Optimal handhold cutout dimensions are
  - 3.8 cm or greater height
  - A minimum length of 11.5 cm
  - 5.0 cm hand clearance
  - A minimum of 0.635 cm wall thickness
  - Semioval in shape
  - A smooth nonslip surface
- Handholds near the bottom of the container allow the worker to carry the load near knuckle height and minimize static muscle loading of the upper extremities
- The edges of the container should be rounded, not sharp, to eliminate risk of contract stress between the load and the hand, arm, and body

To summarize, the factors to consider in evaluating the load; material handling consider the following categories:

- Load: mass or measure of force required, carrying, pushing, or pulling
- Dimensions of container
- Distribution of load
- Quality of coupling
- Stability of load

8.5.2.3 Work Practice Characteristics

In the implementation of risk mitigating interventions, three broad categories to consider include engineering controls, administrative controls, and personal protective equipment (PPE). Applying ergonomic principles to each of these areas is necessary in designing safe MMH tasks.

8.5.2.3.1 Engineering Controls

Engineering controls are preferred over other intervention measures because they address the sources of ergonomic hazards. Engineering controls or designs should reduce the physical exertion or stamina requirements of a task by providing a work system (equipment, tools, furniture, processes, methods, work flow, and environment) that reduces the physical demands on the worker and allows people to safely and effectively perform a job. Engineering controls often use "assistive technologies" or mechanical aids to perform the job. Engineering-based design or redesign of the workplace should seek to eliminate the risk factors associated with chronic and acute musculoskeletal injuries, particularly low back pain. This entails minimizing MMH tasks and personal exposure to excessive loads, bending, twisting, reaching, vibration, and prolonged sitting or standing. The categories that are to be considered in engineering controls for material handling include the following:

- Workplace is adequately sized for task performance, equipment, and movement.
- Materials should not be staged or placed on the floor but at appropriate heights for load handling.
- Conserve momentum wherever possible and let gravity work to your advantage.
- Environmental considerations should include the type of flooring and floor finishes to be used.

8.5.2.3.2 Administrative Controls

The alternative to engineering controls is the use of administrative controls, PPE, or personnel selection. These are inherently less effective than eliminating or significantly reducing the root cause of the hazard through design or redesign. The purpose of administrative controls is to limit the duration of personal exposure to the risk factors associated with MMH tasks. Administrative controls can take many forms including:

- Job rotation (rotating the exposed population into less physically demanding jobs or jobs that do not tax the same muscle groups as the job of concern)
- Job enlargement or enrichment (providing added task variety, adding less taxing aspects to the job, and sharing tasks among several muscle groups)
- Allow job to be self-paced
- Increasing the number of people performing the job (thereby spreading the exposure to a wider population, but reducing individual exposure duration)
- Training in safe handling techniques
- Worker selection and placement

8.5.2.3.3 Personal Protective Equipment

Personal protection that engineering and administrative controls cannot offer may be provided through PPE. PPE is equipment designed to be worn or attached to the body to promote safety, reduce exposure, ease of task performance, or other ergonomic benefits. Desirable qualities of PPE are that they perform the function for which they are intended, properly fit, and do not require the user to exert greater force than is otherwise necessary. Additionally PPE should not require the user assume extreme postures or substantially limit mobility. Typical MMH PPE includes safety shoes, gloves, eye protection, back belts, and hard hats. PPE should be provided in a variety of sizes in order to accommodate anthropometrics of various workers. Workers issued PPE should be trained in how the equipment is to be used, its limitations, care, useful life, and proper disposal. The Occupational Safety and Health Standards for General Industry, sections 1910.132 through 1910.137, provide guidance on PPE (OSHA, 1910.132–137).

8.5.2.4   Worker Characteristics

The goal of ergonomics is to fit the task to the person, rather than identifying people who can perform a job; thus, the evaluation of worker characteristics should not be used to eliminate potential workers. Nonetheless, it is important to understand the level of risk associated with a worker population. This information can guide decision making in task design and support economic justification in instances where task design will reduce the risk for a large portion of the worker population. A variety of individual characteristics have been identified as possible risk factors in the development of low back pain in material handling tasks. These include a history of back injury, poor fitness levels, second jobs, recreational activities, hobbies, smoking, age, gender, obesity, physical stature, and psychosocial issues (including family, financial or other personal difficulties, job or management dissatisfaction, a lack of job control, and work-related stress among other factors) (NC OSHA, 2009). To understand the likelihood for injury, the worker characteristics should be considered in the following categories (the examples provided are not the exhaustive list of factors to be considered within each category):

- *Physical*—age, gender, anthropometry
- *Sensory capabilities*—visual, auditory, or tactile senses
- *Psychomotor*—measure of worker motor skills capability
- *Personality*—risk acceptance level
- *Health status*—previous injuries, fitness level, acclimation to environments
- Leisure time activities

Each of these categories of factors should be considered in assessing overall personal risk of injury in MMH task performance.

8.5.3   Risks of Manual Material Handling Tasks

As previously stated MMH manual is among the most frequent and most severe causes of injury all over the world, often resulting from overexertion. Overexertion involves the lifting of loads that are too heavy or at improper heights. Overexertion injuries of all types in the United States occur at a rate of about 500,000 workers/

year or 1 in 20 workers. Overexertion was claimed as the cause of lower back pain by over 60% of people suffering from it. If overexertion injuries involved low back pain with significant lost time, less than one-third of the patients eventually return to their previous work. Approximately two-thirds of overexertion injury claims involved lifting loads and about 20% involved pushing and pulling loads (OSHA, 2011).

Manipulation of a load, even a light weight, may cause a strain because the body is required to stretch, bend, twist, or straighten using fingers, arms, trunk, and/or legs. The strains can be dynamic or static, with either a fast or slow start, and for short or long durations of time involving a single or several events. Back injuries can be extremely painful and frequently result in long absences from work and early disability.

Additional causes of MMH back injuries are improper lifting, which is lifting done that is not in the sagittal plane, and fatigue, which results from not allowing adequate recovery from exertion. When considering the musculoskeletal structures within the trunk, a variety of elements may be individually or collectively strained. Tension strains can be in the form of linear elongation or of bending movements or twisting torque; thus, shoulders and arms can also be affected by MMH tasks.

MMH problems are generally addressed by evaluations of lifting, pushing or pulling, and slipping risks through physiological and biomechanical applications. The primary focus of these applications for MMH tasks has been the lower back, specifically the disks of the lumbar spine. Nevertheless, this is not the only region of the back that can be negatively affected by inappropriate or excessive MMH; thus, consideration of risk must be given to the entire vertebral column.

8.5.4 ASSESSMENT OF WORKLOAD FOR MANUAL MATERIAL HANDLING TASKS

MMH tasks may involve the exertion of single maximal efforts, intermittent loading or the handling of light loads over long periods of time. Therefore, these tasks commonly require the assessment of the impact of the physical workload. Several assessments of work load exist and are frequently used to measure or predict the energy costs, as well as the human capabilities of various types of tasks. A useful model for assessing impact of the task on a worker should include the following parameters (Mital et al., 1993; OSHA, 2011):

1. Body weight of individuals doing the task (range)
2. Weight of the load
3. Vertical start and end positions of a lift
4. Dimensions of the load
5. Frequency of handling
6. Gender

A number of methods and models exist to assess physical loads associated with MMH. Sophisticated software programs such as MADYMO (Figure 8.3) have the ability to produce complex biomechanical analysis. The most appropriate method will depend on the task characteristics, worker population, potential risk and resources available for the analysis.

8.5.4.1 Biomechanical Models

Biomechanical methods used to evaluate strains and human capabilities have been widely used in the past and in MMH they are mainly focused on the spinal column, specifically on the responses of the vertebrae and the vertebral discs to compression. However, the calculation of compression strain can be difficult because the human spinal cord is not a straight column. The setup of the spinal column provides it with the elasticity it needs in order to absorb the shocks experienced during our daily activities. The spinal column has an elongated "S" shape. A slight backward curve is found at the chest level, called the kyphosis and a slight forward curve is found in the lumbar region, called the lumbar lordosis. Intervertebral discs separate the vertebrae in the spine (Figure 8.2).

Aging and repetitive motions can cause degeneration of the discs, but often it is overexertion that leads to severe injuries. Any sudden compression force on a degenerated disc causes pressure either on the spinal cord itself or on the nerves that run out of the spinal cord, resulting in a herniated disc. The more weight sustained by the upper body, the larger the force on the spinal column.

The vertebra column support the weights of the head and neck, arms and hands, and of the upper trunk, any additional load can further compress the spine. This can be particularly detrimental to the lumbar vertebrae. Also, movements such as walking, running, bending, or twisting increase the force created on the spinal column.

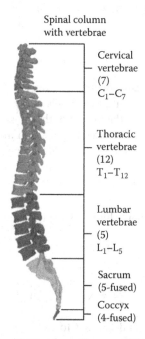

Spinal column
with vertebrae

Cervical vertebrae (7) C_1–C_7

Thoracic vertebrae (12) T_1–T_{12}

Lumbar vertebrae (5) L_1–L_5

Sacrum (5-fused)

Coccyx (4-fused)

FIGURE 8.2 Spinal Column with Vertebrae (Source: Wikipedia, The Free Encyclopedia, accessed August 2011)

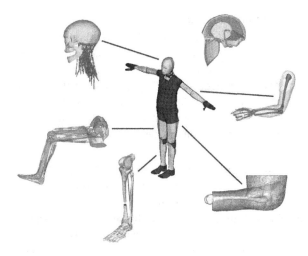

FIGURE 8.3   MADYMO Software Human Male Occupant Crash Safety and Biomechanical Modeling Software   (Source: TASS website http://www.tass-safe.com/en/home; accessed August 2011)

Pressure on the nerves, narrowing of the spaces between vertebrae, and pulling and squeezing of the tissues, muscles, and ligaments at the spine can lead to of discomfort, aches, muscular cramps, and paralyses (i.e., lumbago and sciatica).

Biomechanical modeling can be highly complex using computer systems such as MADYMO (Figure 8.3) or as simple as a traditional manual link analysis (Figure 8.4). The model selected for a biomechanical analysis should be compatible with the evaluation needs and provide sufficient detail to asses impact and associated risks.

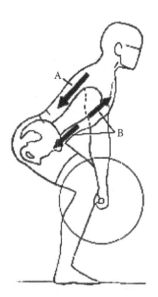

FIGURE 8.4   Intrabdominal Pressure

8.5.4.2 Intra-Abdominal Pressure

Lifting tasks create an increase in pressure between the inferior and superior diaphragms of the trunk cavity due to the contractile motions of the abdominal muscles. At first, it had been assumed that intra-abdominal pressure helped minimize the compression strain experienced in the spine while lifting loads in the hands. However, new research and studies have revealed that the intra-abdominal pressure does not reduce spinal compression loading. Nonetheless, inter-abdominal pressure has been used to evaluate loading on the lower vertebrae.

8.5.4.3 Physiological Approach

MMH abilities can be measured using the physiological approach. This approach is most appropriate for MMH tasks that are done frequently and over some duration of time, for example, an 8 h/day task. Also, this approach is concerned with energy consumption and stresses acting on the cardiovascular system.

Measuring the physical load using a physiological approach involves monitoring the body's response to a load. The measures then physical loading include gas exchange, change in HR, sweating and other. The HR or cardiac activity is considered the most convenient physiological measure of job stress. Increased HRs can be reflected in the stress of various conditions including

- Physical effort
- Environment heat and/or humidity
- Psychological stress and/or time pressure
- Cognitive load: decision making and perpetual work
- Environmental factors: chemicals and noise
- Combinations of any of the factors in the preceding points

Stress disturbs the body's equilibrium, also known as homeostasis. The human body strives to maintain homeostasis when any type of heavy physical exertion takes place.

Homeostasis: According the *American Heritage®* Medical Dictionary, homeostasis is the "tendency of an organism or cell to regulate its internal conditions, such as the chemical composition of its body fluids, so as to maintain health and functioning, regardless of outside conditions" (American Heritage Dictionary, 2011). The organism or cell maintains homeostasis by monitoring its internal conditions and responding appropriately when these conditions deviate from their optimal state. During heavy work, several adaptations take place in the body and these include

- Deeper and more rapid breathing
- Increased HR
  - Increase blood flow to the muscles
- Vasomotor adaptations
  - Dilation of the blood vessels in the organs involved
  - Diversion of blood from the organs not immediately concerned into those that need more oxygen and nutrients

- Rise in blood pressure to speed up the flow of blood
- Increased supply of sugar
- Rise in body temperature and increased metabolism in order to convert more chemical energy into mechanical energy

8.5.4.4 Psychophysical Approach

One of the most widely utilized methods for establishing criteria for MMH tasks is the psychophysical approach (Mital et al., 1993). The premise of this approach is that people integrate and combine both biomechanical and physiological stresses in their subjective evaluation of perceived stress. In this approach, subjects are allowed to adjust a load to the maximum amount that they feel they can sustain without strain or discomfort and without becoming unusually tired or out of breath. The maximum selected is called the maximum acceptable weight limit (MAWL) of the load.

8.6 ASSESSMENT OF ENERGY EXPENDITURES: DIRECT AND INDIRECT MEASURES

Direct measures of energy expenditure can be calculated through computerized measurement of gas. Measurement of oxygen uptake can be computed during any given task. For example, the Douglas bag can be used and it involves the subject inhaling air from the atmosphere and exhaling it through a mask connected by tubing to a large brown bag. After the empty bag is filled with expired air, the test is terminated. The oxygen content of the air in the bag can be compared with that of the atmosphere to determine the amount of oxygen metabolized by the subject. If the time taken to fill the bag is known, the subject's rate of oxygen uptake can be calculated. The energy expenditure may also be measured in the laboratory with a gas exchange measurement technique that captures all expired and inhaled gases (see Figure 8.5).

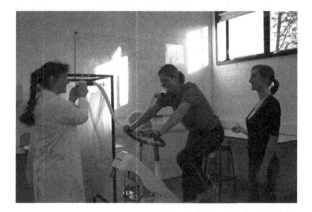

FIGURE 8.5 Gas Exchange (Source: Department of Sport and Exercise at Aberystwyth University.)

Subjective measures of energy expenditure include user response based approaches. The Borg rating of perceived exertion (RPE) is a subjective method that can be used to obtain perceived exertion.

Changes in HRs can either be continuous or intermittent. Following are a few of the definitions for HR values that are used in this type of assessment.

- *Resting pulse*: average HR before the work begins
- *Working pulse*: average HR during the work
- *Work pulse*: difference between the resting and working pulses
- *Total recovery pulse (recovery cost)*: sum of the heart beats from the end of the work until the pulse returns to its resting level
- *Total work pulse (cardiac cost)*: sum of the heart beats from the start of the work until the resting level is restored

Energy required to perform physical work is correlated to the percent heart rate change, which can be estimated by using the following formula:

$$\% \text{ Heart rate change} = \frac{(HR_{task} - HR_{rest})}{(HR_{max} - HR_{rest})} \tag{8.2}$$

where
HR_{rest} is the resting heart rate
HR_{task} is the task heart rate
HR_{max} is the maximum heart rate

8.7 EVALUATION OF PHYSICAL EFFORTS IN MANUAL MATERIAL HANDLING

Estimating energy consumption or task demands is possible by considering the intensity and duration of the work being done. These types of task elements are broken into either primary or secondary physical effort requirements. Primary physical efforts include manual handling tasks, such as lifting, pulling, and climbing. The intensity of task performance is a function of a variety of factors including: muscular requirements, task duration, and mental stress associated with the task performed. Intensity levels are divided into five levels:

Level I: Light intensity/occasional
Level II: Light intensity/frequent moderate intensity/occasional
Level III: Light intensity/constant moderate intensity/frequent heavy intensity/ occasional
Level IV: Heavy intensity/frequent moderate intensity/constant
Level V: Heavy intensity/constant

8.8 NIOSH LIFTING GUIDELINE

In 1981, a panel of experts prepared for the U.S. NIOSH, a work practice guide for manual lifting. This document contained distinct recommendations for acceptable masses to be lifted that differed from the previous assumptions that one could establish one given weight for men, women, or children that would be safe to lift. In this equation, two thresholds were established. The lower, called the action limit (AL), was thought to be safe for 99% of working men and 75% of working women. The actual limit values depend on the starting height of the load, the length of its upward path (vertical distance), the distance in front of the body (horizontal distance), and the frequency of lifting. In places where the existing weight mass was above the actual limit value, engineering or managerial controls had to be applied to bring the load value down to the acceptable limit. The second threshold, called the maximum permissible load (MPL), was considered to be three times the actual limit. Under no circumstances was continuation of the lifting task to be allowed if the existing load was above the MPL, or three times larger than the action limit. The guidelines established during NIOSH 1981 state the following:

1. Under the optimum conditions 40 kg can be lifted
2. The MPL corresponds to the capacity of a 75th percentile male
3. The AL corresponds to the 99th percentile woman
4. MPL = 3(AL)
5. The AL should not exceed 3400 N at the L5/S1
6. Loads exceeding the AL are considered a normal risk
7. Workloads that exceed the MPL are unacceptable
8. Loads between the two limits require training and/or selection of the workforce
9. If possible keep at or below the AL

Flaws were found in the assumptions of the NIOSH 1981 guideline. Therefore, a decade later, in 1991, NIOSH revised the technique for assessing overexertion hazards of manual lifting (NIOSH, 1991). The new document no longer contained two separate weight limits but only one, called the recommended weight limit (RWL). The RWL represents the maximal weight of a load that may be lifted or lowered by about 90% of American industrial workers, male or female, physically fit, and accustomed to physical labor. This new equation resembles the 1981 formula for AL, but includes new multipliers to reflect asymmetry and the quality of hand-load coupling. The 1991 equation allows as maximum a "load constant" (LC), which is permissible under the most favorable circumstances, with a value of 23 kg (51 lb). The RWL is calculated as followed:

$$RWL = LC * HM * VM * DM * AM * FM * CM \qquad (8.3)$$

where
   LC is the load constant of 23 kg or 51 lb
   HM is the horizontal multiplier, where H is the horizontal distance of the hands from the midpoint of the ankles
   VM is the vertical multiplier, where V is the vertical location (height) of the hands above the floor at the start and end points of the lift

DM is the distance multiplier, where D is the vertical travel distance from the start to the end points of the lift

AM is the asymmetric multiplier, where A is the angle of symmetry, such as in the angular displacement of the load from the medial (mid-sagittal plane), which forces the operator to twist the body. It is measured at the start and end points of the lift

FM is the frequency multiplier, where F is the frequency rate of lifting, expressed in lifts per minute

CM is the coupling multiplier, where C indicates the quality of coupling between hand and load

Each multiplier may assume a value between zero and one

Additionally, the lifting index (LI) is calculated by taking the actual weight of the load, divided by the RWL. The LI is the ratio of the load being lifted to the RWL.

$$LI = \frac{\text{Actual weight lifted}}{\text{RWL}} \tag{8.4}$$

The LI is intended to provide a means of comparing lifting tasks. A LI greater than 1.0 is likely to pose an increased risk of low back pain for some fraction of the exposed workforce and may be used to identify potentially hazardous lifting jobs. The higher the LI, the greater the risk to the persons performing the task. The goal should be for all lifting tasks to have a LI less than 1.0.

To calculate the RWL using the NIOSH lifting equation, the following steps should be followed:

- Determine the weight of the load (contents and container) being lifted.
- Assess operator position and lifting task by obtaining the following parameters
    - Horizontal distance (H) of lift
    - Vertical distance (V) of lift
    - Carry distance (D)
    - Frequency of lift (F)
    - Angle of asymmetry (A)
    - Quality of coupling (C)
- Determine the appropriate multiplier factors for each parameter using the NIOSH tables (see NIOSH Lifting Equation Applications Manual), or software tools.
- Calculate the RWL for the task.
- Calculate the LI.
- Compare weight of the load against determined weight limit for the task.
- Determine if the LI exceeds 1, and if there is a risk associated with the task.

The tables associated with this equation are contained in the accompanying lab manual as well as the manual for application of the NIOSH lifting guideline. This

TABLE 8.3

General Design/Redesign Suggestions for NIOSH Lifting Guideline

If HM is less than 1.0	Bring the load closer to the worker by removing any horizontal barriers or reducing the size of the object. Lifts near the floor should be avoided; if unavoidable, the object should fit easily between the legs
If VM is less than 1.0	Raise or lower the origin or destination of the lift. Avoid lifting near the floor or above the shoulders
If DM is less than 1.0	Reduce the vertical distance between the origin and the destination of the lift
If AM is less than 1.0	Move the origin and destination of the lift closer together to reduce the angle of twist, or move the origin and destination further apart to force the worker to turn the feet and step, rather than twist the body
If FM is less than 1.0	Reduce the lifting frequency rate, reduce the lifting duration, or provide longer recovery periods (i.e., light work period)
If CM is less than 1.0	Improve the hand-to-object coupling by providing optimal containers with handles or handhold cutouts, or improve the handholds for irregular objects
If the RWL at the destination is less than at the origin	Eliminate the need for significant control of the object at the destination by redesigning the job or modifying the container or object characteristics

Source: NIOSH, 1991.

document is available at the NIOSH website (Waters et al., 1994). The NIOSH lifting guideline provides detail, tables, examples and additional details on the use of this method. A few general design redesign suggestions given the outcome of the parameters is provided in Table 8.3 (NIOSH, 1991).

8.8.1 OTHER LIFTING GUIDELINES

8.8.1.1 ACGIH Threshold Limit Values for Lifting

The American Conference of Governmental Industrial Hygienists (ACGIH) recommends guidelines for safe lifting. The threshold limit values (TLVs) for lifting recommend upper and lower limits based upon frequency, duration, and other risk factors associated with lifting.

8.8.1.2 University of Michigan 3D Static Strength Prediction Program

The 3D Static Strength Prediction Program software predicts static strength requirements for tasks such as lifts, presses, pushes, and pulls. The program provides an approximate job simulation that includes posture data, force parameters, and male/female anthropometry. The results include the percentage of men and women who have the strength to perform the described job, spinal compression forces, and data comparisons to NIOSH guidelines. The user can analyze torso twists and bends and make complex hand force entries. Analysis is aided by an automatic posture generation feature and 3D human graphic illustrations (Figure 8.6).

3D static strength prediction program (3DSSPP) software

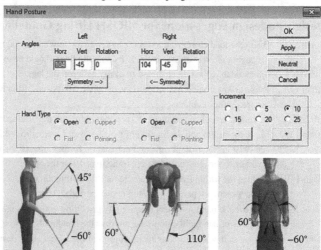

FIGURE 8.6   3D SSPP

8.8.1.3   Ohio State Lumbar Motion Monitor

Traditionally, most workplace ergonomic assessments have focused on joint loading in static postures. However, epidemiologic studies have shown that 3D dynamic motion is associated with an increased risk of occupational injury and illness. The Biodynamics Lab at Ohio State University has developed a unique research program that focuses on the study of occupational joint loading under realistic dynamic motion conditions. The program's goal is to obtain a better understanding of how much exposure to realistic risk factors is too much (http://biodynamics.osu.edu/research.html).

8.8.1.4   Snook's Psychophysical Tables

In 1978, Liberty Mutual (Snook and Cirello, 1978) published extensive tables of loads and forces found acceptable by male and female workers for continuous MMH jobs. This data was first updated in 1983, revised in 1991, and again updated in 1993 by Snook, Cirello, and Hughes (Snook and Cirello, 1991). The following assumptions apply to these guidelines:

- Two-handed symmetrical material handling in the medial (mid-sagittal) plane, in other words directly in front of the body; yet, a light body twist may occur during lifting or lowering
- Moderate width of the load, such as 75 cm (28.5 in.) or less
- A good coupling, for example, of hands with handle and shoes with floor
- Unrestricted working postures

- Favorable physical environment, such as about 21°C at a relative humidity of 45%
- Only minimal other physical work activities
- Material handlers who are physically fit and accustomed to labor

This approach is captured in the Liberty Mutual Lifting Tables. The recommendations in the Liberty Mutual Tables differ from the guidelines established by the NIOSH of 1991 in the additional detail regarding lifting capability. The Liberty Mutual Tables guidelines are grouped with respect to the percentage of the worker population to whom the values are acceptable, while the NIOSH guidelines are not. Also, the Liberty Mutual Tables data are divided by gender, whereas the NIOSH values are unisex. This approach subdivides lifts into three different height areas:

- Floor to knuckle height
- Between knuckle and shoulder heights
- Shoulder to overhead reach

However, it is interesting to note that both the new NIOSH guidelines and the revised data that took place in 1991 by Liberty Mutual indicate that the lack of handles reduces the loads that people are willing to lift and lower by an average of 15% (Snook, 1978; Snook and Cirello, 1991).

Comparison of Lifting Evalutaion Techniques: A study by Russell et al. (2007) compared five lifting analysis tools. The project assessed and compared the results of maximum lifting capability for the following approaches:

- NIOSH Lifting Guideline
- American Conference of Governmental Industrial Hygienist (ACGIH) Threshold Limit Values (TLVs) for Lifting
- Liberty Mutual MMH Tables
- University of Michigan, 3D Static Strength Prediction Program (3DSSPP)
- Washington State Department of Labor and Industries (WA L&I) Lifting Calculation

To enable comparisons between the five lifting assessment instruments, the output of each approach was converted to an exposure index (Table 8.4).

The study results indicate that the NIOSH guideline, ACGIH TLV, and Liberty Mutual instruments provided similar results when assessing musculoskeletal exposures associated with a lifting task. Additionally, the study concluded that when considering the complexity of performing these calculations, the ACGIH TLV, Liberty Mutual, and WA L&I methods were simpler to use since they required fewer inputs than the six inputs needed for the NIOSH lifting equation; however, NIOSH offered a greater range of interpretive capabilities in order to understand what aspects of the lift would benefit most from changes (Russell et al., 2007).

TABLE 8.4

Work Systems Definitions

Shift	The hours of a given day that an individual or a group of individuals is scheduled to be at the workplace.
Off time	The hours of a given day that an individual or a group of individuals is not normally required to be at the workplace.
Schedule	The sequence of consecutive shifts and off time assigned to a particular individual or group of individuals as their usual work assignment.
Permanent hours	A schedule for an individual or a group of individuals that does not normally require them to work more than one type of shift. That is, the time of day one works is constant.
Rotating hours	A schedule that normally requires an individual or a group of individuals to work more than one type of shift. That is, the time of day one works changes.
Basic cycle	The minimum number of days required to complete the specific sequence of shifts and off time constituting a given schedule. The number of days until a schedule begins to repeat.
Major cycle	The minimum number of days required to arrive at a point where the basic sequence of a schedule begins to repeat on the same days of the week. The number of days until the basic sequence falls on the same days when it repeats.
Work system	All of the schedule(s) implemented in a given workplace to meet the real or perceived requirements of a given plant, process, or service.

Source: Gavriel, S. (Ed.): *Handbook of Human Factors and Ergonomics.* 1997. Copyright Wiley-VCH Verlag GmbH & Co. KGaA.

8.8.2 GENERAL ERGONOMIC RULES FOR LIFTING OF LOADS

The following guidelines can be useful for industrial manual handling tasks, particularly when lifting loads. Practicing these rules can help reduce the risks of injury in MMH intensive tasks.

- Reduce the sizes, weights, and forces involved.
- Provide good handholds.
- Keep the object close to the body, always in front and do not twist.
- When lifting, keep the trunk up and the knees bent.
- Minimize the distance through which the object must be moved.
- Move horizontally, not vertically.
- If possible, convert lifting and lowering to pushing, pulling, or carrying (provide material at the proper working height).
- Plan all movements and make them smooth.

8.9 FATIGUE

In order to get a better understanding of the term "fatigue" in the industrial environment, three broad categories that are primary to fatigue in the workplace have been established. These include

1. Muscular fatigue
2. Mental fatigue
3. Shift work related fatigue

8.9.1 Muscular Fatigue

Muscular fatigue is generally manifested in a reduction in muscular power and slower movements. Muscular fatigue can be extremely painful and contributes to reduce capability in task performance. As previously discussed, during muscular contraction chemical reactions take place to provide the energy required to produce mechanical work. After a muscle is contracted, energy reserves need to be replenished. However, if the demand for energy exceeds about half the person's maximal oxygen uptake, lactic acid and potassium ions accumulate, leading to muscle fatigue.

The phenomenon of muscular fatigue is better explained in regard to maintained static muscle contraction. If the muscular effort surpasses about 15% of a maximal voluntary contraction, the blood flow through the muscle is diminished. This blood flow can even be cut off when a maximal effort is exerted. A lack of blood flow causes an accumulation of potassium ions as well as a depletion of sodium in the extracellular fluid. These biochemical events in conjunction with an intracellular buildup of phosphate, caused by the breakdown of ATP, disturb the coupling between nervous excitation and muscle fiber contraction and produce muscle fatigue. The resulting fatigue can be alleviated with several rest periods. If the ratio of "total resting time" to "total working time" is the same, then several short rest periods have a better "recovery value" than many long rest periods. To minimize the amount of muscular fatigue that occurs, it is important to not overexert or exceed muscular duration guidelines in task performance. Also, minimizing any static loading is critical.

8.9.2 Mental Fatigue

Mental fatigue considers the overloading of the sensory organs and mental capacities and can impact the ability to perform physically intensive tasks as well as cognitive tasks. This type of fatigue is the result of excessive cognitive loading, as a result of task performance personal factors and/or environmental factors. Essentially, any task that requires continual mental effort in excess of human capabilities can result in this disorder.

Generally, people start to show signs of mental fatigue when the working day is almost over. During this time, it becomes more difficult to concentrate, tasks appear to be much more complicated, and mistakes are more common. Similarly, late night work or studying can also result in mental fatigue, making it more difficult to retain information and mistakes appropriate cognitive task design, rest and are more frequent. Although, there is no cure for mental fatigue, following a well-balanced and nutritious diet can help improve this condition. Another way to reduce mental fatigue is through exercise. Exercising increases oxygen levels in the bloodstream, which in turn helps improve cognitive processes.

Visual fatigue, eyestrain (asthenopia), eye discomfort, subjective visual symptoms, or distress result from the use of one's eyes (National Research Council Committee on Vision, 1983, p. 153). Near work is thought to cause visual fatigue and ocular symptoms. Temporary shifts in the near point of accommodation and the resting position of convergence can also produce visual fatigue.

Near work is thought to cause visual fatigue and ocular symptoms. Other factors that lead to visual fatigue include vigilant tasks, using viewing equipment (i.e., microscope or binoculars) and frequent accommodation due to changes in viewing distance. Additional factors include

- Temporary shifts in the near point of accommodation and the resting position of convergence
- Recession of the near point suggests a loss of refractive power and is sometimes referred to as accommodation strain
- Rigidity of visually demanding work routines and extended work periods

The occurrence and impact of visual fatigue can be minimized when the task, processes, and equipment are assessed to ensure consistency with human capabilities.

8.9.3 SHIFT WORK RELATED FATIGUE

Studies of rotating shift work and night work indicate that shift work represents a physiological strain on humans (Rodahl, 1989). A normal work shift is generally considered to be a work period of no more than eight consecutive hours during the day, 5 days a week with at least an 8 h rest before returning to work. Any shift that incorporates more continuous hours, requires more consecutive days of work, or requires work during the evening can be considered extended shifts, unusual work hours, or shift work. Extended shifts may be used to maximize scarce resources. Tables 8.4 and 8.5 provide descriptions and definitions of shift work terminology (Tepas et al., 1997).

Shift work as defined is not a recent phenomenon. Night watches have been recorded in the earliest written history and many battles were initiated in early hours to gain advantage over a resting enemy. The industrial revolution, particularly around the 1920s, led to three shift rotations to take advantage of mechanical equipment on a 24 h basis. Additionally, many service industries are expected to provide continual presence in critical tasks such as

- Emergency telephone operators
- Police
- Fire fighters
- Hospital staff
- Military personnel
- Restaurant workers

OSHA does not have a specific regulation addressing shift work schedule, design, or safety but it is governed under the OSH Act's General Duty Clause, Section 5(a)

TABLE 8.5
Work Shift Definitions

Three-Shift Systems

First shift	A work period of about 8 h in duration that generally falls between the hours of 0600 and 1700. Also known as morning or day shift.
Second shift	A work period of about 8 h in duration that generally falls between the hours of 1500 and 0100. Also known as afternoon-evening or swing shift.
Third shift	A work period of about 8 h in duration that generally falls between the hours of 2200 and 0700. Also known as night or graveyard shift.

Two-Shift Systems

Day shift	A work period of about 10 or more hours in duration that generally falls between the hours of 0800 and 2200.
Night shift	A work period of about 10 or more hours in duration that generally falls between the hours of 2200 and 0800.

Other Work Shift Classifications

Split shift	Any work period that is regularly scheduled to include two or more work periods less than 7 h, separated by more than 1 h away from work, on the same day.
Irregular shifts	Work periods that vary their shift starting time and duration in an erratic way.
Non-workday	Any calendar day in which only off time is scheduled.

Source: Gavriel, S. (Ed.): *Handbook of Human Factors and Ergonomics.* 1997. Copyright Wiley-VCH Verlag GmbH & Co. KGaA.

(1) of the OSHA Act. Statistically, accident rates are higher during night shifts, and employees who work long or irregular hours are at greater risk of injury. Shift workers also face a higher risk of stress-related health problems than employees who work regular hours.

8.9.3.1   Organizational Approaches to Shift Work

The most important considerations in shift schedule design are the job performance capabilities of people at different hours, physiological effects of night work, and the psychosocial effects on workers, their families, and friends. The Ergonomics Group of Eastman Kodak (Eastman Kodak, 1983) defined a process for designing shift schedules to accommodate requirements for 5 and 7 day shift schedules that include 8 h, 12 h, and combinations of other work hours. An example of using these guidelines in the development of a work shift for 12 h rotating shifts for 5 day rotations is shown in Table 8.6.

Effective shift schedule design or optimizing the design of the shift schedule is an means to reduce ergonomic, health, and safety problems as well as counteracting some of the social challenges for shift workers. Satisfaction with a particular shift system is the result of a complicated balancing act that is the best compromise for personal, psychological, social, and medical concerns. The Canadian Centre for

TABLE 8.6
Five Day Shift Schedules—12 h Shifts

A Shift

	1	2	3	4	5	6	7	8	9	10	11	12
Employee	W	T	F	S	S	M	T	W	T	F	S	S
A1	D	D	—	—	—	D	D	—	—	D	D	D
A1	D	D	—	—	—	D	D	—	—	D	D	D
A2	N	N	—	—	—	N	N	—	—	N	N	N
A2	N	N	—	—	—	N	N	—	—	N	N	N
Supervisor	12P	12P	—	—	—	12P	12P	—	—	12P	12P	12P

B Shift

	1	2	3	4	5	6	7	8	9	10	11	12
Employee	W	T	F	S	S	M	T	W	T	F	S	S
B1	—	—	D	D	D	—	—	D	D	—	—	—
B1	—	—	D	D	D	—	—	D	D	—	—	—
B2	—	—	N	N	N	—	—	N	N	—	—	—
B2	—	—	N	N	N	—	—	N	N	—	—	—
Supervisor	—	—	12P	12P	12P	—	—	12P	12P	—	—	—

Source: Eastman kodak, 1983.
Notes: D, day shift; N, night shift; day's work week: Saturday 1200 to Saturday 1200; night's work week: Sunday 0000 to Sunday 0000; supervisor works noon to midnight; work week is Saturday to Saturday 1800.

Occupational Health and Safety (CA OSHA, http://www.ccohs.ca, 2011) lists the following factors to consider in shift work design:

- Length of the rotation period (the number of days on any one shift before switching to the next shift). The optimum length of the rotation period has been disputed.
  - The most common system has a rotation period of 1 week, with five to seven consecutive night shifts. However, since it generally takes at least 7 days for adjustment of the circadian rhythms, it is argued that just as adjustment starts to occur, it is time to rotate to the next shift.
  - Some researchers suggest that a longer shift rotation should be arranged so that the worker spends from 2 weeks to 1 month on the same shift thus allowing circadian rhythms to adjust. A problem occurs when the worker reverts to a "normal" day/night schedule on days off, thus, possibly canceling any adaptation. Also, longer periods of social isolation may result.
  - Others suggest a rapid shift rotation where different shifts are worked every 2–3 days. This system may reduce disruption to body rhythms because the readjustment of circadian rhythms is minimized. It also provides time for some social interaction each week.
  - Individual differences and preferences also play an important role.

- Direction of rotation of shifts. Shifts should rotate forward from day to afternoon to night because circadian rhythms adjust better when moving ahead than back.
- Start and finish times. Early morning shifts are associated with shorter sleep and greater fatigue. Avoid shift start times as early as 5:00 a.m. when possible. Finish times should be at times that are consistent with standard shifts ending. In other words, finish times for late shifts are desirable at hours that will allow the worker to still obtain a reasonable amount of nighttime rest (i.e., do not end a shift at 3:00 a.m.).
- Length of rest between shifts. Allow a rest period of at least 24 h after each set of night shifts. The more consecutive nights worked, the more rest time should be allowed before the next rotation occurs.
- Alternative forms of organizing work schedules.
  - Extended work days of 10 or 12 h have been used. Due to the advantage of fewer consecutive night shifts and longer blocks of time off.
  - Additional fatigue from long work hours may also have adverse effects on shift work.
  - The physical and mental load of the task should be considered when selecting the length of a work shift.
  - Ergonomic hazards and exposure to chemical or physical agents should also be considered when selecting a shift system.

8.9.3.2 Additional Considerations

- Provide time off at "socially advantageous" times like weekends whenever possible.
- Start a special shift system if production demands result in extended periods of overtime work.
- Inform shift workers of their work schedules well ahead of time so they and their families and friends have sufficient time to adapt and plan.
- Allow as much flexibility as possible for shift changes.
- Keep schedules as simple and predictable as possible.

Fatigue from shift work often results from the individual's inability to rest during off hours and the body's natural tendency is to slow down in the evening and night hours. The circadian rhythm affects the bodily functions by signaling these functions to slow down at night and increase in the day. Circadian rhythms are physical, mental, and behavioral changes that follow a roughly 24 h cycle, responding primarily to light and darkness in an organism's environment. This rhythm is found in most living things, including animals, plants, and many tiny microbes. The study of circadian rhythms is called chronobiology. The discussion of circadian rhythm in shift work is primarily concerned with alertness for task performance and the ability to rest in off hours. The amount and quality of daytime sleep is less than nighttime sleep. The general consensus is that shift work is less productive and that the workers tend to have difficulty adjusting. In fact, the circadian clock can be reversed after several days of shift work, but this change takes an enormous amount of time to change permanently. Other issues with shift work include social contacts that do not change just because you work at night, the inability to get enough quality sleep, and knowledge of clock time. According to OSHA (2010), fatigue is common in shift work and employers should be mindful of the symptoms. The symptoms of fatigue, both mental and physical, vary and depend on the person and his or her degree of overexertion. Some examples include

- Weariness
- Sleepiness
- Irritability
- Reduced alertness, lack of concentration and memory
- Lack of motivation
- Increased susceptibility to illness
- Depression
- Headache
- Giddiness
- Loss of appetite and digestive problems

In the design of tasks, schedules, and expectations for shift work, consideration should be given to ergonomic and human factors guidelines to reduce the likelihood for fatigue, enhance safety, and enhance productivity.

8.9.4 General Fatigue

The degree of fatigue depends on different stresses endured throughout the work-shift. In order to recuperate, rest periods are necessary, especially during nighttime. General fatigue symptoms can be subjective or objective and the most common symptoms are the following:

- Subjective feelings of weariness, somnolence, faintness, and distaste for work
- Sluggish thinking
- Reduced alertness
- Poor and slow perception
- Unwillingness to work
- Decline in both physical and mental performance

8.9.5 Measuring Fatigue

It is very valuable for ergonomists to assess levels fatigue associated with a task. This is essential to understand the relationship that exists between fatigue, work output, and level of stress. No singular method of directly measuring the extent of fatigue itself exists. However, studies have been performed to measure "certain manifestations or indicators of fatigue." Some of the methods that have been used in studies include

1. Quality and quantity of work performed.
   a. Indirect way of measuring fatigue
   b. Quantity of output expressed as time taken to make one unit, number of items processed or fabricated, or number of operations performed per unit time
   c. Quality of the output is also measured, for example, mistakes and faulty items
2. Recording of subjective perceptions of fatigue: questionnaires are used to measure subjective feelings.
3. *Electroencephalography (EEG)*: increase in alpha and theta rhythms and reduction of beta waves used as indicators of weariness and sleepiness.
4. Measuring frequency of flicker-fusion of eyes.
   a. Subject exposed to flickering lamp
   b. Flickering rate is increased until it appears as a continuous light
   c. Lowering of flicker-fusion rate is a sign of fatigue
   d. Flicker-fusion technique has been used less in recent years
   e. Controversial results
   f. Quantitative measure of fatigue are difficult
   g. Rare correlations with other symptoms of fatigue
5. *Psychomotor tests*: measure functions that involve perception, motor reactions, and interpretation.
   a. Typical tests include
      i. Simple and selective reaction times
      ii. Tests involving touching or pricking squares in a grid
      iii. Tests of skill
      iv. Driving tests under simulated conditions
      v. Typing
   b. Decrease in performance interpreted as increase in fatigue
   c. Signs of fatigue may be inaccurate due to workers' excitement or interest in the test
6. Mental tests.
   a. Includes tests such as arithmetic problems and estimation, concentration, and memory tests
   b. Decrease in performance interpreted as increase in fatigue
   c. Signs of fatigue may be inaccurate due to workers excitement or interest in the test

The level of fatigue is based on the measurements obtained from the methods and these measurements are usually taken before, during, and after a task is executed. It is important to remember that the results are only relatively significant since the values need to be compared with someone who is not experiencing any type of stress.

In addition to the methods previously mentioned, other types of fatigue exist including general bodily fatigue, nervous fatigue, and chronic fatigue. General bodily fatigue results from physically overloading the entire body. Nervous fatigue is caused by overstressing one part of the psychomotor system, for example, as in skilled and often repetitive work. Chronic fatigue is a result of an accumulation of long-term effects of physical and psychological factors and can impact task performance and overall individual well-being. This condition may need to be treated by a medical professional.

8.10 SUMMARY

Although, the mechanization of machines in many industries has helped reduce the need for workers to perform strength-intensive tasks, there are still many organizations that require very heavy jobs. Therefore, the human body needs to be able to produce enough energy to perform work. The respiratory, circulatory, and metabolic systems play a significant role in the production of energy for work performance. People have different muscle strengths, which depend on the subject's age, gender, health, training, as well as other factors. However, more importantly, the tasks should be designed to be compatible with worker capabilities, MMH, such as lifting, can result in severe injuries of the lower back. Prevention of unnecessary strains, over exertions, and other MMH injuries is essential for ethical occupational and economic reasons. Proper lifting guidelines and detailed recommendations for appropriate handling of loads have been established making task requiring handling of materials easier and less vulnerable to injuries.

Case Study

Article—Case Study #4: Shift Work

Problems with 12 Hr Nursing Shifts; Pharmacy Ergonomics; Older Workers; Multi-touch Interfaces

By Peter Budnick

June 24, 2010

Study Reveals Widespread Fatigue, Risk for Errors With 12-h Nursing Shifts

A common practice of successive 12 h shifts for U.S. hospital nurses leaves many with serious sleep deprivation, higher risk of health problems, and more odds of making patient errors, according to a University of Maryland, Baltimore (UMB) study presented today at the 24th annual meeting of the Associated Professional Sleep Societies in San Antonio.

The 12 h shift trend started in the 1970s and 1980s when there were nursing shortages, said Jeanne Geiger-Brown, PhD, RN, associate professor with the School of Nursing at UMB. Hospitals started giving nurses more benefits and bonuses, eventually leading to emphasis on 12 h shifts, negotiated by the nursing profession, while hospitals saw that the change made nurses happy and bought into it, she said.

"Nurses often prefer working a bunch of 12-hour shifts and then lots of time off. But, I contend that it is not a good thing for nurse planning," said Geiger-Brown. The study involved 80 registered nurses, working three successive 12-hour shifts, either day or night. "We were surprised at the short duration of sleep that nurses achieve between 12-hour shifts. Over 50 percent of shifts were longer than 12.5 hours, and with long commutes and family responsibilities, nurses have very little opportunity to rest between shifts.

The study also found that the average total sleep time between 12 h shifts was only 5.5 h. Night-shift nurses averaged only about 5.2 h of sleep, and the quality of their sleep was extremely fragmented. People who are sleep deprived experience micro-sleep periods, little lapses in attention, and intershift fatigue, "meaning that on the next shift you don't fully recover from the previous one," said Geiger-Brown.

Geiger-Brown was the co-author of a review article in the March issue of *The Journal of Nursing Administration*, "Is it Time to Pull the Plug on 12-Hour Shifts?" It analyzed evidence from several recent scientific studies of the safety risks involved with long work hours, and challenges the current scheduling paradigm. "Few hospitals offer alternatives to the pattern," Geiger-Brown commented. "There is increasing evidence that 12-hour shifts adversely affect performance. In 10 previously published studies of the effects of 12-hour shifts, none showed positive effects, while four showed negative effects on performance."

Most recent studies cited in the article point to an increase in patient care errors related to successive 12 h shifts. Geiger-Brown cites one study of 393 nurses on 5317 shifts who were surveyed anonymously. The odds of making errors by those who reported working more than 12 h in shifts was three times greater than nurses who reported working 8.5 h shifts. Experiencing partial sleep deprivation chronically, over many years, is dangerous to the nurses' health and to the patients. The most common problems with an over emphasis on 12 h shifts are needle-stick injuries, musculoskeletal disorders, drowsy driving, and other health breakdowns related to sleep deprivation.

The authors don't expect 12 h shifts to end anytime soon. However, there are several "tools" hospitals can deploy to help nurses and hospital administrators better manage the practice, such as courses in harm reduction, fatigue risk management, and more training of nurses about the risks.

Posting Date: 06/08/2010

Contact Name: Steve Berberich

Contact Phone: 410-706-0023
Contact Email: sberb001@umaryland.edu
Source: Total contents of Case Study reprinted with permission from Ergoweb.com

EXERCISES

8.1 Define heavy work and industry and task characteristics that can lead to this type of work.

8.2 List the specific data necessary to perform a NIOSH load-lifting analysis.

8.3 What are the job or task conditions not included in the NIOSH analysis?

8.4 Perform a lifting analysis of an actual occupation task using the Liberty Mutual Tables?

8.5 What are the primary risk factors associated with MMH work?

8.6 What are the primary types of occupational fatigue? Describe each and provide an ergonomics approach to mitigate the risk of each type of fatigue.

8.7 Identify a federal guideline, industry guideline, or ISO standard that offers ergonomics guidance in the design of material handling tasks.

   a. Explain what type of occupational ergonomic concerns in shift work. What four different types of literature and data were consulted by NIOSH experts in assembling the NIOSH *Guide for Manual Lifting*?

REFERENCES

Bridger, R.S. (2008). *Introduction to Ergonomics*, 3rd edn., Chapter 8, CRC Press: New York, pp. 309–314.

Canadian Centre for Occupational Health and Safety. (2011). Rotational shiftwork. http://www.ccohs.ca/oshanswers/ergonomics/shiftwrk.html (accessed September 2011).

Cheung, Z., Hight, R., Jackson, K., Patel, J., and Wagner, F. (2007). *Ergonomic Guidelines for Manual Material Handling*, DHHS Publication 2007-131, National Institute for Occupational Safety and Health: Atlanta GA. Retrieved October 7, 2008.

Coffiner, B. (1997). Ergonomic design of the snow shovel. Report Occupational Health Clinics for Ontario Workers, Inc.

Department of Labor Occupational Safety and Health Program, Cherie Berry Commissioner of Labor. (2009). http://www.nclabor.com/osha/etta/indguide/ig26.pdf (accessed September 2011).

ErgoWeb. Retrieved February 1, 2011, from http://www.ergoweb.com/news/detail.cfm?id=566

Exxon Valdez Oil Spill Trustee Council. (1990). Details about the accident. Retrieved February 1, 2011, from http://www.evostc.state.ak.us/facts/details.cfm

Freivalds, A. (1986a). The ergonomics of shoveling and shoveling design—A review of the literature. *Ergonomics*, 29(1), 3–18.

Freivalds, A. (1986b). The ergonomics of shoveling and shoveling design—A review of the literature. *Ergonomics*, 29(1), 19–30.

Gavriel, S. (Ed.) (1997). *Handbook of Human Factors and Ergonomics*, Wiley Interscience: New York.

Grandjean, E. (1988). *Fitting the Task to the Man*, Taylor & Francis: New York.

Hettinger, T. and Muller, E.A. (1953). Muscle strength and training. *Arbeitsphysiologie*, 15, 111–126.

Kodak, E. (1983). *Ergonomic Design for People at Work—Volume 1*, Van Nostrand Reinhold Company Inc.: New York.

Kodak, E. (1986). *Ergonomic Design for People at Work—Volume 2*, Van Nostrand Reinhold Company Inc.: New York.

Kroemer, K.H.E. and Grandjean, E. (1997). *Fitting the Task to the Human: A Textbook of Occupational Ergonomics*, 5th edn., Taylor & Francis: London, U.K.

Lehmann, G. (1953). *Praktische Arbeitphysiologie*, Thiem Verlag: Stuttgart, Germany (chapter titled Published Ergonomics Literature by T. Megaw contained in International encyclopedia of ergonomics and human factors, Volume 3 edited by W. Karwowski, Taylor & Francis, Boca Raton, FL, 2006).

Mital, A., Nicholson, A.S., and Ayoub, M.M. (1993). *A Guide to Manual Materials Handling*, Taylor & Francis: London, U.K.

National Research Council, Video Displays, Work and Vision. (1983). *Commission on Behavioral and Social Science and Education (CBASSE)*, National Academy Press: Washington, DC.

NC OSHA, A Guide to Manual Materials Handling and Back Safety. North Carolina Department of Labor Occupational Safety and Health Program, Cherie Berry Commissioner of Labor. (2009), Retrieved September 2011, from http://www.nclabor.com/osha/etta/ind-guide/ig26.pdf

NIOSH. (1991). *Work Practice Guide for Manual Lifting*, U.S. Department of Health and Human Services: Cincinnati, OH.

NIOSH. (1995). *A Strategy for Industrial Power Hand Tool Ergonomics Research—Design, Selection, Installation, and Use in Automotive Manufacturing*, U.S. Department of Health and Human Services: Cincinnati, OH, pp. 25–31.

NIOSH. (1997a). *Elements of Ergonomics Programs*, U.S. Department of Health and Human Services: Cincinnati, OH, Publication No. 97-117.

NIOSH. (1997b). *Musculoskeletal Disorders and Workplace Factors: A Critical Review of Epidemiologic Evidence for Work-Related Musculoskeletal Disorders of the Neck, Upper Extremity, and Low Back* (Ed. B. Bernard), Department of Health and Human Services, Public Health Service, Centers for Disease Control and Prevention, National Institute for Occupational Safety and Health: Atlanta, GA, DHHS (NIOSH) Publication No. 97-141.

NIOSH. National occupational research agenda for musculoskeletal disorders research topics for the next decade. A Report by the NORA Musculoskeletal Disorders Team.

Orwell, G. (1961). *Down and Out in Paris and London: A Novel*, Harcourt Brace Jovanovich: New York, p. 91.

Rodahl, K. (1989). *The Physiology of Work*, Taylor & Francis: London, U.K.

Russell, S.J., Winnemuller, L., Camp, J.E., and Johnson, P.W. (2007). Comparing the results of five lifting analysis tools. *Appl. Ergon.*, 38, 91–97.

Sanders, M.M. and McCormick, E.J. (1993). *Human Factors in Engineering & Design*, 7th edn., Chapter 8, McGraw-Hill: New York, pp. 245–247.

Snook, S.H. (1978). The design of manual handling tasks. *Ergonomics*, 21(12), 963–985.

Snook, S.H. and Ciriello, V.M. (1991). The design of manual handling tasks: Revised tables of maximum acceptable weight and forces. *Ergonomics*, 34(9), 1197–1213.

Taylor, F.W. (1913). *The Principles of Scientific Management*, Harper & Bro: New York.

Tepas, D., Paley, M., and Popkin, S. (1997). Work schedules and sustained performance. *Handbook of Human Factors and Ergonomics*, Chapter 32, John Wiley & Sons: New York, pp. 1021–1058.

Washington State Legislature. (1985). Flexible time work schedules. http://apps.leg.wa.gov/RCW/default.aspx?cite=41.04.390

Waters, T.R., Putz-Anderson, V., and Garg, A. (1994). *Applications Manual for the Revised NIOSH Lifting Equation*, U.S. Department of Health and Human Services, Centers for Disease Control: Cincinnati, OH, DHHS (NIOSH) Publication No. 94-110.

9 Information Ergonomics, Controls, and Displays

9.1 LEARNING GOALS

The learning goals of this chapter include teaching the concepts of information processing, measurement of mental workload, and an introduction to ergonomic design of controls and displays. The student will be able to perform a subjective analysis of mental workload upon understanding the material contained within this chapter.

9.2 KEY TOPICS

- Information processing
- Information processing models
- Cognition and ergonomics
- Information workload measurement
- Controls and displays

9.3 INTRODUCTION AND BACKGROUND

Mental workload can be viewed as the perception of effort or demands experienced in the performance of a cognitive task. The amount of mental workload varies with the individual's tolerance and is influenced by the amount of work, the time given to perform the task(s), the cognitive demands, and the subjective perception of the difficulty of the task(s) (Cain, 2007).

9.4 INFORMATION PROCESSING

When information is received, it must be compared, combined with data already stored in the brain and put into memory in its new form. Factors that can affect this process include experience, knowledge, mental agility, and the ability to think thoroughly to generate new ideas.

The interaction between a human and a machine relies on the exchange of information between the operator and the system. Detailed and precise models of machines and machine behavior have been made available by designers in order to improve the interaction between humans and machines. Information can be thought of as the transfer of energy that has meaningful implication in any given situation.

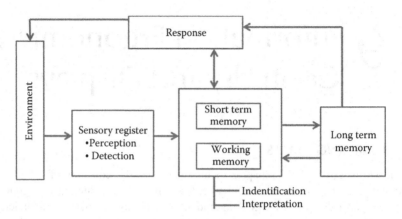

FIGURE 9.1   Components of human memory. (Adapted from Gagne and Driscoll, 1988.)

Information can be defined as the absorption of interpreted data, directly or indirectly, through human sensory functions. The unit of measurement that is used in measuring information is the bit. Information is quantified in bits (binary digits), and represented by H. A bit can be thought of as an amount of information necessary to decide between two equally likely alternatives and is derived with the following formula:

$$H = \log 2n$$

where n is the number of equally probable alternatives.

Thus, the more alternatives in the decision, the more information needed to make the decision.

Many models of information processing exist (NASA, 2011; Newell and Simon, 1972; Sternberg, 1977) and the basic components of these models are

- Sensory receipt (register)
- Short-term memory (also referred to as working memory, however, there are differences)
- Long-term memory
- Response
- Reaction

The components of human memory are shown in Figure 9.1.

9.4.1 Perception

Perception is the process of attaining awareness or understanding of how sensory information or sensation occurs. Perceptual processing has two important features. First, it generally proceeds automatically and rapidly, requiring little attention. Second, it is driven both by sensory input, as the senses receive the information that is to be perceived. Perception is partially determined by an analysis of the stimulus or environmental input and relayed from the sensory receptors by the sensory or "lower" channels of neural information. This aspect of processing is referred to as bottom-up perceptual processing (Pastorino and Doyle-Portillo, 2006). On the other hand, when sensory evidence is poor, perception may be driven heavily by our expectations based on past experience, and produce top-down processing (Pastorino and Doyle-Portillo, 2006). Bottom-up and top-down processing usually work together, supporting rapid and accurate perceptual work. Varying levels of perception exist and depend on the stimulus and the task.

The most primitive form of perception is detection. Detection simply involves knowing whether or not a signal or target is present. Signal detection theory is one of the primary means used to evaluate perception. It is usually used when there are two possible states, for example, an audible symbol is either detectable or not. It is also a good approach to use when evaluating a person's detection of a particular stimulus for an emergency situation. The outcomes of the signal detection theory are summarized in the following and shown in Table 9.1:

In signal detection analyses, the signal is present (or not presented) and the subject is ask to state whether or not a signal was detected. The potential outcomes are below.

- Hit
  - Saying there is a signal when there is a signal.
- False alarm
  - Saying there is a signal when there is no signal.
- Miss
  - Saying there is no signal when there is a signal.
- Correct rejection
  - Saying there is no signal when there is no signal.

9.4.2 HUMAN AS INFORMATION PROCESSOR

Numerous theories of information processing exist in an attempt to describe the flow of information. The following are commonly included in information processing: perception, coding and decoding, learning, memory and recall, reasoning, making judgments and decisions, transmitting information, and executing the desired physical responses (Figure 9.2). The theories of information processing fall into two general categories: serial and parallel. Serial models view information processing as a set of discrete sequential stages, whereas parallel (also known as distributed or connectionist) models describe information processing as occurring in multiple locations across a network of connections. Three of the models that are used to represent or measure information processing to represent or measure. These models can present serial or parallel constructs of information processing they include.

- *Mathematical models*: model the amount of information processed.
- *Physical models*: attempt to model and explain the physical processes that take place in information processing.
- *Structural models*: list the major components of information processing.

TABLE 9.1

Reference for Some of the More Common Workload Assessment Techniques

Types of Measures/Techniques	References
Subjective	
Bedford	Roscoe (1987), Roscoe and Ellis (1990)
Modified Harper–Cooper	Wierwilli and Gasali (1983)
NASA-TLX	Hart and Staveland (1988)
Psychophysical scaling	Gopher and Braune (1984)
SWAT	Reid and Nygren (1988)
WORD	Vidulich et al. (1991)
Workload profile	Tsang and Velazquez (1996)
Performance	
AGARD STRESS battery	AGARD (1989)
Choice reaction time	Kalsbeek and Sykes (1967), Krol (1971)
Criterion task set	Shingledecker (1984)
Multi-attribute task battery	Comstock and Arnegard (1992)
Time estimation/interval production	Wierwille and Connor (1983)
Mental arithmetic	Brown and Poulton (1961), Harms (1986)
Sternberg memory	Wickens et al. (1986)
Tracking task	Levison (1979), Wickens (1986)
Psychophysiological EEG/ERPS	*Biological Psychology* (1995), 40 (1 & 2)
Eye blinks	Stern and Dunham (1990)
Heart rate variability	*Biological Psychology* (1992), 34 (2 & 3)
Respiration	*Biological Psychology* (1992), 34 (2 & 3)
Analytic	
PROCRU	Baron and Corker (1988)
Queuing theory	Moray et al. (1991)
Timeline analysis	Kirwin and Ainsworth (1992)
TAWL	Hamilton, et al. (1991)
TLAP	Parks and Boucek (1989)
VCAP	McCracken and Aldrich (1984)
W/INDEX	North and Riley (1989)

Source: Gavriel, S. (Ed.): *Handbook of Human Factors and Ergonomics*. 1997. Copyright Wiley-VCH Verlag GmbH & Co. KGaA.

9.4.2.1 Serial Models

The "stage theory" of information processing, developed by Atkinson and Shiffrin (1968), describes three stages for storing and processing information. The first stage is the sensory memory in which stimuli from the environment are received and stored very briefly. In the second stage, short-term memory (or working memory), information is temporarily stored and processed just long enough to act upon the information. Information in short-term memory is either discarded or moved to

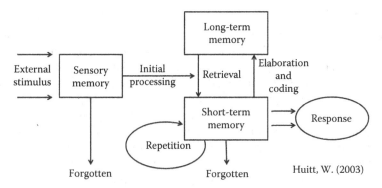

FIGURE 9.2 Information processing.

the third stage, long-term memory (Huitt, 2003). Jarvis (1999) describes Welford's (1968) three-stage information processing model as similar to that of a computer processing operation. In the first stage, perception involves receiving information (or inputs). The second stage results in decision making because at this point the information is processed. This processed information then results in the third stage, response and the associated output Figure 9.3.

Similar to Welford's model is Whiting's three-stage model that views sensory inputs received through perceptual mechanisms, then information processed and decisions made through translator mechanisms, and finally actions taken by effector mechanisms (Whiting's Model; Mac, 2010).

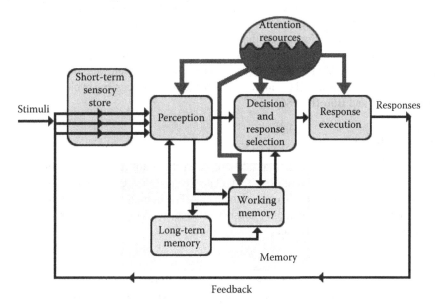

FIGURE 9.3 Information processing model and response execution. (Source: adapted from Wickens, C.D., Engineering Psychology and Human Performance, Harper Collins, New York, 1992. Accessed at Federal Aviation Administration Website (http://www.hf.faa.gov/webtraining/Cognition); accessed September 2011.)

9.4.2.2 Parallel Models

In contrast to the serial models that view information processing as occurring in a series of logical steps, parallel-distributed processing models view information processing as occurring simultaneously through multiple interactions occurring across the brain's neural network. Rumelhart and McClelland (1986) explain that "information processing takes place through the interaction of a large number of simple processing elements." Parallel-distributed models offer numerous advantages over serial models including similarity to the physiological structure of the brain and better explanation of processing multiple, simultaneous, and conflicting information, which would be too cumbersome under the serial approach (Rumelhart, 1986). An individual may process information using both conceptual approaches, alternating between methods based on the level of focus or attention required for varying components of the task performance.

9.4.3 Channel Capacity Theory

The channel capacity theory states that a certain quantity of information is delivered to the input end of the channel and the output at the other end depends upon the capacity of the channel. When the input is small, all the information is transmitted through the channel. However, if the input increases, it eventually reaches a threshold value where the output from their channel is no longer a representation of all information transmitted. Every day, humans experience incoming information that exceeds the channel capacity of their information processing capability. Therefore, only a small portion of the information is absorbed and processed by the brain. Humans throughout the information procssing cycle have an extensive channel capacity for information that is communicated through the spoken word; however, multiple sensory inputs must also be registered and processed. The decision is made by the human as to which information is most relevant and the priority of processing is determined throughout the information processing cycle (Figure 9.3) (Schweickert and Boroff, 1986).

9.4.4 Memory

Memory refers to the process that takes place in the brain that involves the storing of incoming information, retaining it, and later retrieving it. Often, only a fraction of the incoming information is stored in the brain. The human memory can be categorized into three aspects including sensory storage, working memory or short-term memory, and long-term memory.

9.4.4.1 Sensory Memory

Each sensory channel appears to have a temporary storage mechanism that prolongs the stimulus representation for a short time after the stimulus has ceased. The visual and auditory sensory channels are most often referenced. The visual storage system, called the iconic storage, lasts for less than 1 s before fading away, while the auditory storage, called the echoic storage, may last for a few seconds before it dissipates.

9.4.4.2   Short-Term (Working) Memory

Working memory or short-term memory (STM) is usually defined as having three core components (Baddeley, 1995): verbal, phonological, and spatial. The verbal component consists of two subsystems: the phonological store and the articulatory loop. The phonological store represents information in linguistic form, typically as words and sounds. The information can be rehearsed by articulating those words and sounds, either vocally or subvocally, using an articulatory loop. The spatial component represents information in an analog, spatial form, often typical of visual images (Logie, 1995). The third component is the central executive, which controls working memory activity and assigns attentional resources to the other subsystems. Previous findings indicate that phonological STM capacity supports learning of novel phonological forms, such as new vocabulary (e.g., Baddeley et al., 1998). It was hypothesized that semantic STM capacity would support the learning of novel semantic information (Freedman and Martin, 2001).

Short-term memory is associated with a temporary storage of information in the brain, where only a small fraction of information is briefly retained while performing a specific mental or physical task. An example of short-term memory is remembering a phone number while writing it down. Since short-term memory has a limited storage capacity, errors are very possible, when excessive demands are placed on the STM. It is believed that short-term memory is coded with three types of codes. These include the visual, which are visual representations of the stimuli; phonetic, which are audible representations of stimuli; and semantic, which include abstract representations of the meaning of the stimuli.

In order for information to be retained by working memory it must be rehearsed. The number of items that can be held in the working memory has been quantified by Miller's chunking theory (Miller, 1956). Miller observed that although some human cognitive tasks fit the model of a "channel capacity" characterized by a roughly constant capacity in bits, short-term memory did not. A variety of studies could be summarized by saying that short-term memory had a capacity of about "seven plus-or-minus two" *chunks*. A chunk is considered a "meaningful" individual bit of information. Miller's chunking theory states that seven plus-or-minus two chunks of information or five to nine pieces of information is the limit of the working memory. Miller's chunking theory STM limitation $S = 7 \pm 2$. This theory can be used as a rule of thumb in designing working memory requirements in task performance.

9.4.4.3   Long-Term Memory

There are three main activities related to long-term memory: storage, deletion, and retrieval. Information is stored or transferred from working memory to long-term memory by semantically coding it. Semantic coding implies that the information is related to information that is already stored in long-term memory and thus attaches significant permanent meaning. Once information is encoded into long-term memory either through learning and training, its long-term representation can take on a variety of forms. Some knowledge is procedural, such as how to do things, and other knowledge is declarative, such as knowledge of facts. Another distinction made is between semantic memory, which is general knowledge of things, such as word meaning and episodic memory, which involves memory for specific events. Long-term memory is extremely stable and can withstand brain disturbances and electric shocks.

Deletion is primarily the result of decay and interference. However, emotional factors, physical trauma, and intense emotional trauma also affect long-term memory. Some studies suggest that information may never be deleted but rather is difficult to retrieve due to the lack of use and interference from other information stored in the long-term memory.

The two categories of long-term memory retrieval include recall and recognition. Recall describes a situation where information that is stored in long- or short-term memory is accessed. An example of this could be trying to recall one's home address. Recognition describes a situation where you are provided with information and you must indicate whether that information corresponds with what is already stored in memory. Failures of recognition and recall often lead to human performance errors. A primary source of forgetting involves the absence of external retrieval cues, to help activate information stored in long-term memory. Another source of forgetting is the passage of time or lack of accessing the information over an extended period of time. Humans tend to remember best those events that have happened most recently, a phenomenon referred to as "recency."

9.4.5 ATTENTION

Attention is the activity that facilitates the processing of specific information present in our surroundings. Four basic types of attention exist and include (Pashler, 1997)

- Sustained attention
- Selective attention
- Focused attention
- Divided attention

9.4.5.1 Sustained Attention

Sustained attention or vigilance requires users to maintain focused attention for extended periods of time while remaining alert. This type of task can be demanding mentally and lead to fatigue if human factors guidelines are not implemented to manage the mental workload. Jobs such as extended driving or flying require sustained alertness, which can be mentally demanding. Guidelines for sustained attention tasks include the following:

- Provide appropriate work-rest schedules and task variation.
- Increase the conspicuity of signals by making them louder, bigger, or more intense.
- Reduce uncertainty as to when or where a signal will likely occur.
- Inject artificial signals and provide operators with feedback on their detection performance.
- Provide training and motivate users.
- Monitor environmental factors.

9.4.5.2 Selective Attention

Selective attention requires the monitoring of several channels or sources of information to perform a simple task. It is up to the person to select between the channels. An example of selective attention is listening to a particular instrument when a full orchestra is performing. Following are a few recommendations when performing selective attention tasks:

- When multiple channels must be scanned for signals, use as few channels as possible.
- Provide information to the user as to the relative importance of the various channels.
- Reduce overall level of stress on the user.
- Provide the user with preview information as to where signals will occur in the future.
- Train the user in effective scanning.

9.4.5.3 Focused Attention

Focused attention requires maintaining attention on one or a few channels of information while not getting distracted by other channels of information. Situations where there are competing channels can occur however these task require an ability to focus on select channels. An example of focused attention is concentrating on a particular feature of a task such as using a securing machine. A few guidelines when carrying out focused attention tasks are summarized in the following text:

- Make the competing channels as distinct as possible from the channel to which the user is to attend.
- Separate in physical space the competing channels from the channel of interest.
- Reduce the number of competing channels.
- Make the channel of interest larger, brighter, louder, or more centrally located than the competing channels.

9.4.5.4 Divided Attention

Divided attention requires the allocation of resources to a variety of task elements. It is important to consider various input modalities for divided attention tasks, for example, auditory signals for one task and visual signals for another. An example of divided attention is reading a book while at the same time watching television. Divided attention does not necessarily mean performing two or more tasks simultaneously as the activities are generally alternated to allow attention to the given channel. Some guidelines for divided attention tasks include the following:

- Whenever possible, the number of potential sources of information should be minimized.
- If time share is likely to stress a person's capacity, it is important to provide relative priorities of the tasks to optimize the strategy of dividing attention.
- Tasks should be dissimilar.
- Training the user for manual task component will minimize interruptions with mental tasks.

9.4.6 Stimuli Response

The speed of reaction has been of great importance to both psychologists and ergonomists. Psychologists use the study of reaction times for insight into mental problems, while ergonomists use it to measure the ability of a person to perform psycho motor tasks. Reaction time refers to the time that elapses from the appearance of a proximal stimulus (for example, light to the beginning of an effectors action, such as a foot or hand movement). Additional time is required for any movement, which is called motion time.

There are two types of reaction times: simple and choice reaction times. Simple reaction time refers to the case where a person knows a particular stimulus will occur and only one reaction is expected. Under optimal conditions, simple visual, auditor, and tactile reaction times are about 0.2 s. Choice reaction times are considered either when one stimulus out of several possible stimuli occurs or when the person has to choose among several possible reactions. Choice reaction time is a logarithmic function of the number of alternative stimuli and responses and is calculated by the following formula (Hicks, 1952):

$$RT = a + b \log_2 N \quad \text{(Figure 9.2)}$$

where
  RT is the reaction time
  a and b are constants dependant on the situation
  N is the number of choices. N can also be replaced by the probability of any particular alternative, $p = 1/N$, becoming the Hick Hyman

Motion time follows reaction time and is the initiation of the designed muscular activity. Movements can either be simple or very complex and are considered to be a function of the distance covered and the precision required. Many well-designed and documented studies have been conducted on movement times. Paul Fitts' performed significant studies, which have become classics (Fitts, 1952). Around the 1950s, Fitts' studies produced important findings in terms of motion time that are still used today. Fitts' demonstrated the following two points:

- When a target precision or width of target was fixed, the movement time increased with the logarithm of distance.
- When the distance was fixed, the movement time increased with the logarithm of the reciprocal of target width.

Therefore, distance and width (target precision) almost exactly compensated for each other. These findings are expressed in the following equation known as Fitts' law.

$$MT = a + b\log_2\left(\frac{2D}{W}\right) \quad \text{(Figure 9.3)}$$

where
  MT is the movement time
  D is the distance covered by the movement
  W is the width of the target

The constants a and b depend on the situation, such as body parts involved, masses moved, tools or equipment used, on the number of repetitive movements, and on training.

The expression $\log_2 (2D/W)$ is often referred to as the index of difficulty

This estimate of movement time can be used to estimate the level of repetition in a task, design processes, and establish task requirements.

Response time is the time that passes between a stimulus and a response to it and is essentially the reaction time plus the motion time. There are several factors that can affect the response time and they include the following:

* *Auditory*: tones that are similar to the stimulus produce slower reaction time.
* *Visual*: dependent on intensity, flash duration, location, and size of the stimulus.
* *Tactile*: similar for different parts of the body. However, the difference in response is a function of movement time.
* *Uncertainty*: the more uncertainty the longer it takes for a decision.
* *Number of choices*: the more choices, the longer the time to make a decision.
* *Aging*: slower for people less than 15 and greater than 60 years.

9.5   MENTAL WORKLOAD MEASUREMENT

Assuming that mental capacity is a limited resource, workload is then defined as the portion of resources expended in performing a given task. Different models of workload measurement exist. De Waard (1996) performed a study that evaluated drivers' mental workload as other secondary tasks to the primary driving task are performed. Mental workload can be determined using the approaches summarized as follows:

* Measures of primary and secondary task performance
* Objective physiological measurements
* Psychophysical assessment
* Subjective assessment approaches

Some of the more commonly used methods for workload assessment and relative references are listed in Table 9.1 (Tsang and Wilson, 1997).

9.5.1   MEASURES OF PRIMARY AND SECONDARY TASK PERFORMANCE

Primary task performance evaluates a person's performance level in the primary task and presumes that zero and full capacities are known. As workload increases, performance changes measurably. It uses secondary, non-obtrusive measures, such as clutter, queue lengths, and error rates. Secondary task performance is used with the intent to access the spare capacity that remains after the allocation of capacity resources to the primary task. If the person allocates some of the resources truly needed for the primary task, then the task performance in the primary task will suffer.

9.5.2 PHYSIOLOGICAL MEASURES

Physiological measures related to level of mental workload include factors such as heart rate, eye movement, pupil diameter, muscle tension, electroencephalogram (EEG), electrooculogram (EOG), and electrocardiogram (ECG).

Relating the level of mental stress to these measures assumes that the individual's level of arousal will vary as a function of workload or stress. Generally, these measures are less convenient to collect and interpret because some of the physiological measures may not be exclusively affected by task performance or mental workload. However, these measures can provide useful additional information (e.g., whether increasing workload produces physiological evidence of stress) and can supplement other workload assessment methods.

9.5.3 PSYCHOPHYSICAL ASSESSMENT

Psychophysics is the study of the relationship between the physical qualities of a stimulus and the perception of those qualities by an individual (Stevens, 1974). The psychophysical approach to workload assessment operates on the premise that humans are able to perceive the strain generated in the body by a given exertion (or work) and can make relative judgments regarding the level of perceived effort (Kroemer et al., 1994). The psychophysical scaling technique allows an individual to assign a value to the perceived magnitude of the exerted effort (mental or physical) (Sherehiy and Karwowski, 2006).

9.5.4 SUBJECTIVE WORKLOAD ASSESSMENT

Subjective assessments rely on questionnaires and feedback from employees to determine perceived workload. If they are taken after the task is complete, they are not considered real-time evaluations and if they are taken during the task, the tests might interfere with the task and increase workload. A summary of some of these subjective approaches is provided in Table 9.2 (Tsang and Wilson, 1997).

Each approach has benefits and the details of the evaluation should be considered when deciding on an assessment approach. Three commonly used subjective workload assessment measurement approaches in human factors include the following:

- NASA task load index (TLX)
- Subjective workload assessment technique (SWAT)
- Modified Cooper–Harper scale

TABLE 9.2
Subjective Assessment Approaches

Subjective Rating Instruments	Approaches		
	Unidimensional vs. Multidimensional	Absolute vs. Relative	Immediate vs. Retrospective
Bedford	U	A	I
MCH	U	A	I or R
Psychophysical	U	RS	R
SWORD	U	RR	R
NASA-TLX	M	A	I
SWAT	M	A	I
Workload profile	M	A	R

Source: Gavriel, S. (Ed.): *Handbook of Human Factors and Ergonomics.*
1997. Copyright Wiley-VCH Verlag GmbH & Co. KGaA.

Notes: MCH, Modified Cooper–Harper; U, unidimensional; M, multidimensional; A, absolute rating; RS, relative rating to a single standard; RR, redundant relative rating; I, immediate; R, retrospective.

9.5.4.1 NASA Task Load Index

The NASA-TLX is a multidimensional subjective workload rating technique that derives an overall workload score based on a weighted average of ratings on six subscales (Hart and Staveland, 1988). These subscales include mental demands, physical demands, temporal demands, performance, effort, and frustration (Hart, 2006).

The overall workload score is based on the weighted average of rating on the six workload dimensions (Sherehiy and Karwowski, 2006). The NASA-Task Load Index (TLX) can be used to assess workload in various human-machine environments such as aircraft cockpits; command, control, and communication (C3) workstations; supervisory and process control environments; and simulations and laboratory tests. Software has been developed to apply this approach to the measurement of mental workload (Cao et al., 2009) and is available as a free product (Table 9.3).

9.5.4.2 Subjective Workload Assessment Technique (SWAT)

The SWAT is based on the additive multidimensional model of mental workload (Reid and Nygren, 1988). SWAT has been used in a number of studies and measures workload in three dimensions including time load, mental effort load, and psychological stress load, where each of these factors is measured on a scale with three levels. Subjects are asked to rank the 27 possible combinations of levels on the three scales before providing ratings for particular tasks or events. SWAT has been used in diverse environments, for example: high-G centrifuge; command, control, and communications centers; nuclear power plants; domed flight simulators; task simulators; real flight; and laboratory settings (Potter and Bressler, 1994).

TABLE 9.3

NASA-TLX Rating Scale Definitions

Workload Component	Endpoints	Description
Mental demand	Low/high	How much mental and perceptual activity is required (e.g., thinking, deciding, calculating, remembering, looking, searching, etc.)? Was the task easy or demanding, simple or complex, exacting or forgiving?
Physical demand	Low/high	How much physical activity is required (e.g., pulling, pushing, turning, controlling, activating, etc)? Was the task easy or demanding, slow or brisk, slack or strenuous, restful or laborious?
Temporal demand	Low/high	Did you feel the time pressure due to the rate or pace at which the task or task elements occurred? Was the pace slow and leisurely or rapid and frantic or somewhere in between?
Effort	Low/high	How hard did you have to work mentally and physically to accomplish your level of task performance?
Frustration level	Low/high	How insecure, discouraged, irritated, stressed, and annoyed versus secure, gratified, content, relaxed, and complacent did you feel during the task?
Performance	Good to poor	How successful do you think you were in accomplishing the goals of the task set by the experimenter (or yourself)? How satisfied were you with your performance in accomplishing these goals?

Source: Taken directly from Hart, NASA AMES Research Center, 2007.

9.5.4.3 Modified Cooper and Harper Scale

The Modified Cooper and Harper (MCH) scale was developed for workload assessment in systems in which the task is primarily cognitive, rather than motor or psycho motor intensive. The MCH combines a decision tree and a 10-point scale, where 1 means easy workload, 5 means moderately objectionable, and 10 indicates that the task is impossible to successfully perform (Wierwille and Casali, 1983).

9.5.4.4 Selection of a Subjective Assessment Tool

Subjective measures are becoming an increasingly important tool in system evaluations and have been used extensively to assess operator workload. These approaches have proven to be reliable predictors of mental workload and are consistently used due to their practical advantages (i.e., ease of implementation, nonintrusiveness) and history of providing accurate and sensitive measures of operator load. According to Eggemeir et al. (1991), the suitability of the procedure for the evaluation of mental workload should be evaluated using the following requirements:

- *Sensitivity*: the power of the measure to detect changes in task difficulty or demands.
- *Diagnosticity*: the ability of a tool to identify changes in workload variation and also provide insight into the reason for those changes.
- *Selectivity/validity*: the index must be sensitive only to differences in cognitive demands, not to changes in other variables such as physical workload or emotional stress, not necessarily associated with mental workload for the given task.
- *Intrusiveness*: the measure should not intrude or interfere with the primary task performance or activity that is the object of evaluation.
- *Reliability*: the measure must consistently and reliable measure the mental workload.
- *Implementation requirements*: easily implementable from the standpoint of time, instruments, and software for the collection and analysis of data.
- *Subject acceptability*: subject's perception of the validity and usefulness of the procedure.

9.6 CONTROLS AND DISPLAYS

The siginificance of human factors based design is demonstrated by considering the history of control and display design in military aircraft. During the first 2 years the United States was involved in World War II, there were over 2000 major multiengine aircraft accidents. A laboratory study was performed by U.S. Air Force human factors researchers to determine the best combination of control shapes to use in aircraft crew stations for the various flight functions to facilitate the identification of a given control and discriminate it from the others as well as consistency across aircraft. The results of this study led to the standardization of aircraft research in the late 1940s and 1950s to the identification of the instruments most critical to flight; this optimal standardized arrangement of controls remains in use today. The implementation of study findings in aircraft control design resulted in a significant reduction in system-induced pilot errors and improvements in aviation safety as well as establishing a foundation for research and design on controls.

9.6.1 CONTROLS

Controls, which are called activators in ISO standards, transmit inputs to a piece of equipment. These controls can either be activated by hand or foot and the results of the control inputs are revealed to the operator by displays, indicators, or by the subsequent actions of the machine. The functions of controls include activating or shutting down a machine, discrete setting or a distinct adjustment, for example, selecting a TV channel or entering data via a keyboard; quantitative settings, such as adjusting the temperature on a thermostat, and continuous control, such as steering a car. The type of control selected should be compatible with stereotypical or population expectations. For example, a button and not a lever should be selected for a horn. The size and motion of the control should be compatible with stereotypical experience and past practice. For instance, a steering wheel is preferred over two levers for a car control. Also, direction of operation should be compatible with the stereotypical expectation, such as in a throttle where pushing forward should increase speed while pulling back should slow down the speed (Kroemer et al., 1994). Finally, fine control and small force operations should be done by hand while less fine and large force operations should be performed by the feet. Safety should always be built into the design and maintenance of controls.

9.6.2 Guidelines for Control Layout and Design

The following are several guidelines used in the arrangement and grouping of controls (Kroemer et al., 1994):

1. Locate for ease of operation
   a. Controls should be oriented with respect to operator
   b. If operator has different positions, controls should move with operator
   c. In each position, the arrangement of controls should be the same for all operators
2. Primary controls first
   a. Most important, most frequently used controls should have best positions (ease of operation and reach)
3. Group-related controls together
   a. Controls should be grouped by
      i. Controls that have sequential relations
      ii. Controls related to a particular function
      iii. Functions that are operated together
   b. In each functional group, controls should be arranged by operational importance and sequence
4. Arrange for sequential operation
   a. If controls are operated in a given pattern, they should be arranged accordingly
   b. Common arrangement in Western stereotype: left to right and top to bottom
5. Be consistent
   a. Arrangement of functionally identical or similar controls should be consistent from one panel to another
   b. Consistency in use of colors, shapes and layout should follow industry standards, user expectations and human factor guidelines
6. Dead man control
   a. In the event that the operator becomes incapacitated or lets go of or holds on to the control, the system should be designed to shut down
   b. Should be included in any control that has a risk of substantial injury if activated when operator is incapacitated
7. Guard against accidental activation
   a. Avoid inadvertent activation
   b. Can be achieved with shields or requiring substantial forces or torque for critical activation
8. Pack tightly, but do not crowd

The force or torque applied by an operator to actuate a control should be kept as low as possible especially if the control must be repeated often. However, some measure of resistance to provide feedback and to reduce uncontrolled activation, vibrations, or jerks is important. The force exerted by an operator depends on the control's mechanical built-in resistance and this resistance should be a function of the type of control.

Continuous controls are selected if the operation to be controlled can be anywhere within an adjustment range of the control and it is not necessary to set it in any specific position (i.e., volume control on a stereo). Detent (i.e., gear shift in a car) controls are used for discrete settings, in which the control comes to a rest.

The relationships between the action of the control and the resulting effect should be apparent through similarity, habitual use, proximity, coding, grouping, labeling, and consistency with user expectations. Specific applications require certain types of controls.

9.6.3 TYPES OF CONTROLS

Summarized in the following are several discrete and continuous types of controls. The first eight consist of detent controls and the remaining six are continuous controls (Figure 9.4) (Bullinger et al., 1997; Hedge, 2010).

1. *Key lock or key operated switches*: used to prevent unauthorized operation.
2. *Bar knob or rotary selector switches*: used for discrete functions when two or more positions are necessary.

Path of C. motion	Control	Dimension (mm)	Force F (N) Moment M (Nm)		2 Positions	>2 Positions	Continuous adjustment	Precise adjustment	Quick adjustment	Large force application	Tactile feedback	Setting visible	Accidental actuation
Turning movement	Hand wheel	D: 160–800 d : 30–40	**D** 160–200 mm 200–250 mm	**M** 2–40 Nm 4–60 Nm	◐	◐	●	●	◐	●	○	○	◐
	Crank	Hand (finger) r : <250 (<100) I: 100 (30) d : 32 (16)	**R** <100 mm 100–250 mm	**M** 0.6–3.0 Nm 5–14 Nm	◐	◐	●	●	◐	●	◐	◐	○
	Rotary knob	Hand (finger) D: 25–100 (15–25) h: >20 (>15)	**D** 15–25 mm 25–100 mm	**M** 0.02–0.05 Nm 0.3–0.7 Nm	◐	◐	●	●	◐	○	○	○	◐
	Rotary selector switch	I: 30–70 h: >20 b: 10–25	**D** 30 mm 30–70 mm	**M** 0.1–0.3 Nm 0.3–0.6 Nm	●	●	◐	◐	◐	◐	●	●	○
	Thumbwheel	b: > 8	0.4–5N		◐	◐	●	●	◐	○	○	○	◐
	Rollball	D: 60–120	0.4–5N		○	○	●	●	◐	○	○	○	◐

Legend: ○ Not suitable   ◐ Acceptable   ● Recommended

FIGURE 9.4 Hand- and foot-operated control devices and their operational characteristics and control functions.

3. *Detent thumbwheel*: used if the operation requires a compact input device for discrete steps.
4. *Push button*: used for single switching between two conditions, for the entry of a discrete control order, or for the release of a locking system. Can be used for momentary or sustained contact, with or without detent.
5. *Push-pull switch*: used for discrete settings, particularly on and off.
6. *Legend switch*: appropriate for displaying qualitative information on the status of the equipment, which requires the operator's attention and action.
7. *Toggle switch*: can be used if two discrete positions are required.
8. *Rocker switch*: can be used if two discrete positions are required. They project less from the panel than toggle switches.
9. *Knob or round knobs*: used when little force is required and when precise adjustments of a continuous variable are necessary.
10. *Crank:* used mainly if the control must be rotated many times and requires a moderate degree of force.
11. *Hand wheel*: for continuous nominal two-handed operation and can be used when the breakout or rotational forces are too large to be overcome with a one-handed control.
12. *Lever*: used when a large force or displacement is required for continuous adjustments or when multidimensional movements are required.
13. *Continuous thumbwheel*: used as an alternative to round knobs if the compactness of the thumbwheel is helpful.
14. *Slide*: used to make continuous settings.

Foot-operated controls should be used when there are only two discrete conditions, either on or off, when high forces are involved, or when little accuracy is required. Nevertheless, foot controls are still used for fine, continuous control, which is not recommended.

9.6.4 CODING OF CONTROLS

Controls can be coded simply by adding features that make them more easily distinguished. Multiple types of coding can be implemented in control to design to strongly differentiate controls where multiple control mechanisms exist. There are six primary types of control coding defined (Sanders and McCormick, 1993) in human factors which include

1. Location
   a. This is the most powerful principle. Controls with similar locations should be in the same relative location.
   b. For example, depending on the country making the vehicle, locations of many controls are standardized thus meeting driver expectation of consistent control placement.
2. Color
   a. Items are colored based on the function and the task; however, this coding approach is only effective in well-illuminated environments.

b. Color coding requires a longer reaction time than location coding, since it is first necessary interpret the meaning of the color before the task can be executed.

c. If colors have stereotypical means, it is important to be consistent with this expectation in the use of the color. For example, it has become common to make emergency controls red.

3. Size

a. Size can be a useful coding option when distinguishing between different controls.

b. In order to prevent excessive mental demands, not more than three sizes of controls are recommended.

c. Controls with the same function on different items or equipment should have the same size.

4. Shape

a. Shapes can be useful in coding, especially when the control shape resembles the control function.

5. Labeling

a. A label can be used to describe a control, however, just as with color coding sufficient illumination is necessary to benefit from labeling.

b. Labels can be placed above, underneath, or on top of the control. The location should be clearly visible and the wording reads from left to right (Chapanis and Kinkade, 1972).

c. Vertical labels should not be used since they take longer to read, need to be visible with proper illumination, and should not be covered. Symbols are better than text.

6. Mode of operation

a. Each control should have a unique or different feel, which is unique to the method of operation.

b. Consistent, with population stereotype is necessary.

Coding methods can be combined; however, a combination of different codes can generate a new perception or interpretation for the type set of coding. Coding is also useful for offering redundancy, for example, a traffic light uses color and light position codes to reinforce the meaning of each signal in the display.

9.6.5 Emergency Controls

Emergency situations are inherently stressful and workers may respond by forgetting details or panicking; thus, sound ergonomic design in emergency controls is critical. Therefore, it is important to pay particular attention in the design and location of emergency controls. Several design recommendations are summarized in the following text.

Emergency controls must be well designed to allow for action without errors. Emergency control design recommendations include the following:

- Position emergency controls away from frequently used controls
- Make controls easy to reach

- Make controls large
- Require moderate force
- Make controls easy to activate
- Color emergency controls red

Emergency controls used in industrial settings should include a dead-man's switch, which turns a machine off if a worker is not actively pressing the activation switch. The system has a sensor that will automatically to shut off the system. For example, many industrial machines are designed with dead-man switches that automatically stop when signaled by a trip cord or indication of emergency conditions.

9.6.6 DISPLAYS

Displays are extremely important because they provide the operator with vital information on the status of the work environment, vehicle, or equipment. Displays can be visual such as a scale, light, counter, video display terminal (VDT) or flat panel, auditory for instance, a horn, bell, beep, or recorded voice, and tactile such as Braille writing or shaped knobs. The most appropriate type of display depends on task requirements, operator needs, and expectations. The following are factors that should be applied when using displays:

- Present information in a manner that is easily perceived and interpreted.
- Only display information that is necessary for task performance.
- Display information only as accurately as is required for operator's decisions and control actions.
- Present information in a direct, simple, understandable, and usable form.
- Information must be presented in such a way that failure or malfunction of the display will be immediately obvious.

9.6.7 TYPES OF DISPLAYS

As mentioned earlier, displays can be visual, auditory, or tactile. Auditory displays are suitable in situations where the environment lacks sufficient illumination and the operator must move around. The messages in these displays must be short and simple, require immediate attention, be easily differentiated, and deal with temporal urgency. Auditory displays can be single tones, sounds, or spoken messages. Tones should be at least 10 dB louder than ambient noise. Signal frequencies should be different from those that are dominant in the background noise. Signal frequency should range from 200 to 5000 Hz, with the best range being between 500 and 1500 Hz. Low frequencies should be used for long distance sound dissemination and frequencies lower than 500 Hz should be used when signals must bend around or pass through sound barriers. Also, tonal sounds can be more conspicuous by increasing intensity, by interrupting its repeat pattern, or by changing its frequency.

A visual display is appropriate when there is a noisy environment and the operator is stationary relative to the display. The messages in visual displays should be designed such that they are easily perceived and understood. The four broad categories of visual displays include (Hedge, 2010; Kroemer et al., 1994)

1. Check display
   a. Indicates whether or not a given condition exists.
   b. Generally does not provide extensive detail.
2. Qualitative display
   a. Approximate value.
   b. Trend of a changing variable.
3. Quantitative display
   a. Indicates the exact numeral value that must be read.
   b. Exact information that must be determined.
4. Electronic or video display
   a. Capable of offering a variety of static or dynamic information.
   b. Information can be presented in multiple formats including any combination of check, qualitative, quantitative, video, or other types of visual presentations.

9.6.7.1 Guidance on Color Coding in Displays

When using colors in displays, the following guidelines apply:

- Limit the number of colors to four if the users are inexperienced or the display is infrequent.
- Do not exceed seven colors.
- The colors used should be widely separated from one another in wavelength.
- Do not use various shades of the same color.

Suggested combinations of colors include the following:

- Green, yellow, orange, red, and white
- Blue, cyan, green, yellow, and white
- Cyan, green, yellow, orange, and white

9.6.7.2 Emergency Displays

Emergency displays or signals are more successful when followed by an auditory signal or flashing light. These displays should be larger than general status indicators and the luminance contrast to immediate background should be at least 3–1. Flash rates of emergency displays should be 3–5 pulses per second and if the flasher device fails, the light should remain on steadily. Finally, a warning message such as "danger" or "system failure" should be displayed.

9.6.7.3 Electronic Displays

During the 1980s, electronic displays started to replace mechanical displays. Electronic displays were excessively complex, colorful, dynamic, and hard to interpret. On the other hand, when done properly using human factors considerations, they can be more effective, accurate, and faster in communicating information to an operator. In the occupational environment, there are some instances where these displays required consistent focusing and close attention; this type of display should be used only in an environment where the operator can give the required level of focus to the display without inhibiting task performance or safety. Some essential aspects of electronic displays include

- Geometry
  - The size and viewing angle
- Spatial qualities
  - Imaging capacity, raster modulation, pixel faults, and distortion
- Temporal qualities
  - Image formation time, display time, and flicker
- Photometry and colorimetry
  - Illuminance, contrast, and color

9.6.7.4 Technology Impact on Control and Display Design

New technologies in control display design have made major improvements. Such technologies have created a new generation of thin display and backlight products illuminated by light emitting diodes (LED) lighting and touch screens have added flexibility to the workplace and commercial products. Some of these products include

- Flat screens monitors
- Handheld displays such as personal digital assistants (PDA)
- Electronic "books"
- Tactile displays
- Global positioning satellite (GPS) navigational displays
- Virtual reality display/head-mounted displays (HMD)

9.6.7.5 Touch Screens

Touch screen technology has come full force, especially with the introduction of smart phones, iPods, and other Apple products. Touch sensor technology has enabled users in many different aspects of everyday life, from assistive technology to retail settings to entertainment. These devices are continually being added to multiple environments and can be useful as they eliminate the need for a control or activation device. More research is needed on the long-term ergonomic implications of using touch screens on long-term bases.

9.7 SUMMARY

Obtaining an understanding of the principles of information processing, mental workload, and control/display design guidelines is essential in ergonomic applications for the development of human-machine systems. Applying these guidelines will support the development of systems that are compatible with operator capability, reduce the likelihood for error, and improve efficiency.

Case Study

*Ergoweb® Case Study—**Low-Force Activation Buttons***

Longmate, Arthur R., 1996, Johnson & Johnson, Ergonomic Control Measures in the Health Care Industry, Occupational Ergonomics.

Task Prior to Abatement (Description)

Workers had to fully depress the welder activation buttons at the welding station, which were located in the front plate of the welder base. The task includes getting the assembled stapler from a tray and putting it into the welder nest. Then the worker had to fully depress and hold the cycle activation buttons with both hands until contact of the vibratory horn. Finally, the worker had to get the instrument from the nest and put it into a finished instrument tray.

Task Prior to Abatement (Method Which Identified Hazard)

Increasing medical cases of thumb tendinitis among workers.

Ergonomic Risk Factor (Force)

Workers had to apply 5 lb of force with their thumbs to fully depress the welder activation button. More than 5 lb was actually applied by workers to overcome the 5 lb force quickly.

Ergonomic Solution (Engineering Controls)

1. A modified push button switch was developed. The main compression spring was removed in order to reduce the required activation force to approximately 1 lb. An additional spring was built into the contact set in order to provide sufficient force for returning the button to its original position.
2. A standard low-force push button switch requiring 2 lb of finger or thumb force was developed for application of two contact sets.

Ergonomic Solution (Benefits)

* All workers that perform this task now have reduced exposure to thumb tendinitis.
* There was an 80% reduction in required force for one contact set.
* There was a 60% reduction in required force for two contact set.

Ergonomic Solution (Method Which Verified Effectiveness)

Reduction in medical cases of thumb tendinitis.

Comments

Before implementing the new push button design, the company experienced several options including presence-sensing, captive-coupled activation buttons, and photo-electric light beam sensor activation controls to address the injury issues. There were several undesirable ergonomic considerations and cost issues which came up but were not practical.

Source: Total contents of Case Study reprinted with permission from Ergoweb.com

EXERCISES

9.1 Describe the elements of the human information processing model presented in this chapter.
9.2 Explain signal detection theory and the potential outcomes.
9.3 Describe the different types of attention.
9.4 Explain Miller's chunking theory.
9.5 Explain the NASA TLX approach to subjective workload assessment.

REFERENCES

Atkinson, R.C. and Shiffrin, R.M. (1968). Human memory: A proposed system and its control processes. In: K.W. Spence and J.T. Spence (eds.), *Psychology of Learning and Motivation: II*, Academic Press: New York, p. 249.
Baddeley, A. and Logie, R. (1999). Working memory: The multiple component model. In: A. Miyake and P. Shah (eds.), *Models of Working Memory*, Cambridge University Press: New York, pp. 28–61.
Baddeley, A., Gathercole, S., and Papagno, C. (1998). The phonological loop as a language learning device. *Psychol. Rev.*, 105, 158–173.
Bullinger, H., Kern, P., and Braun, M. (1997). Controls, *Handbook of Human Factors and Ergonomics*, Chapter 21, John Wiley & Sons: New York, pp. 697–728.
Cain, B. (2007). Review of the mental workload literature. Report #RTO-TR-HFM-121-Part-II. Defense Research and Development Canada: Toronto, Ontario, Canada.
Cao, A., Chintamani, K., Pandya, A., and Ellis, R. (2009). NASA TLX: Software for assessing subjective mental workload. *J. Behav. Res. Methods*, 41(1), 113–117.
Chapanis, A. and Kinkade, R. (1972). Design of controls. In: H. Van Cott and R. Kinkade (eds.), *Human Engineering Guide to Equipment Design*, Revised Edition, U.S. Govt. Printing Office: Washington, DC, pp. 345–379.
De Waard, D. (1996). The measurement of drivers' mental workload, PhD thesis, Traffic Research Centre, University of Groningen, Haren, the Netherlands.
Eggemeier, F.T., Wilson, G.F., et al. (1991). Workload assessment in multi-task environments. In: D.L. Damos (ed.), *Multiple Task Performance*, Taylor & Francis, Ltd.: London, U.K., pp. 207–216.
Fitts, P.M. (1954). The information capacity of the human motor system in controlling the amplitude of movement. *J. Exp. Psychol.*, 47(6), 381–391. (Reprinted in *J. Exp. Psychol.: General*, 121(3), 262–269, 1992.)
Freedman, M. and Martin, R. (2001). Dissociable components of short-term memory and their relation to long-term learning. *Cognit. Neuropsychol.*, 18, 193–226.
Gavriel, S. (Ed.) (1997). *Handbook of Human Factors and Ergonomics*, 2nd edn., Wiley-Interscience: New York.
Hart, S.G. and Staveland, L.E. (1988). Development of NASA-TLX (Task Load Index): Results of empirical and theoretical research. In: P.A.M. Hancock and N. Meshkati (eds.), *Human Mental Workload*, North-Holland: Amsterdam, the Netherlands, pp. 139–183.

Hart, S.G. (2006). NASA-Task Load Index (NASA-TLX); 20 years later, *Proceedings of the Human Factors and Ergonomics Society 50th Annual Meeting*, HFES: Santa Monica, CA, pp. 904–908.

Hedge, A. (2010). DEA 3250/6510 Class Notes, Cornell University: Ithaca, NY.

Hick, W.E. (1952). On the Rate of Gain of Information. *Q. J. Exp. Psychol.*, 4, 11–26.

Huitt, W. (2003). The information processing approach to cognition. *Educational Psychology Interactive*, Valdosta State University: Valdosta, GA. Retrieved [date] from, http://www.edpsycinteractive.org/topics/cognition/infoproc.html

Jarvis, M. (1999). *Sport Psychology*, Routledge Publishing: New York.

Kroemer, K.H.E., Kroemer, H.J., and Kroemer-Elbert, K.E. (1994). *Ergonomics: How to Design for Ease and Efficiency*, Prentice Hall: Englewood Cliffs, NJ.

Leiden, K., Laughery, K., Keller, J., French, J., Warwick, W., and Wood, S. (2001). A review of human performance models for the prediction of human error. (Technical Report.) Prepared for National Aeronautics and Space Administration System-Wide Accident Prevention Program. Ames Research Center: Moffett Field, CA. Micro Analysis and Design, Inc.: Boulder, CO.

Logie, R.H. (1995). *Visuo-Spatial Working Memory*, Erlbaum: Hove, England.

Macmillan, N.A. and Creelman, C.D. (2005). *Detection Theory: A User's Guide*, 2nd edn., Erlbaum: Mahwah, NJ.

Miller, G.A. (1956). The magical number seven, plus or minus two: Some limits on our capacity for processing information. *Psychol. Rev.*, 63, 81–97.

Newell, A. and Simon H. (1972). *Human Problem Solving*, Prentice-Hall: Englewood Cliffs, NJ.

Pashler, H. (1997). *The Psychology of Attention. Bradford Books*, MIT Press: Cambridge, MA.

Pastorino, E.E. and Doyle-Portillo, S.M. (2006). *What Is Psychology?* Thomson/Wadsworth: Belmont, CA.

Potter, S. and Bressler, J. (1994). *Subjective Workload Assessment Technique (SWAT): A User's Guide*, Systems Research Lab: Dayton, OH.

Reid, G.B. and Nygren, T.E. (1988). The subjective workload assessment technique: A scaling procedure for measuring mental workload. In: P.A.M. Hancock and N. Meshkati (eds.), *Human Mental Workload*, Elsevier Science Publishers B.V. (North-Holland): Amsterdam, the Netherlands, pp. 185–218.

Rumelhart, D.E., McClelland, J.L., and the PDP Research Group. (eds.) (1986). *Parallel Distributed Processing: Explorations in the Microstructure of Cognition*, Volume I, MIT Press: Cambridge, MA.

Sanders, M.S., and McCormick, E.J. (1993). *Human Factors in Engineering and Design*, 7th edn., McGraw-Hill: New York. http://books.google.com/books?id=1bK_LSLD9C4C

Schweickert, R. and Boruff, B. (1986). Short-term memory capacity: Magic number or magic spell? *J. Exp. Psychol.: Learn. Mem. Cogn.*, 12, 419–425.

Sherehiy, B. and Karwowski, W. (2006). Knowledge management for occupational safety, health and ergonomics, *Hum. Factors Ergon. Manuf.*, 16(3), 309–320.

Sternberg, R.J. (1977). *Intelligence, Information Processing, and Analogical Reasoning: The Componential Analysis of Human Abilities*, Erlbaum: Hillsdale, NJ.

Tsang, P. and Wilson, G. (1997). Mental workload, *Handbook of Human Factors and Ergonomics*, Chapter 13, John Wiley & Sons: New York, pp. 417–449.

Welford, A.T. (1968). *The Fundamentals of Skill*, Methuen: London, U.K.

Wierwille, W.W. and Casali, J.G. (1983). A validated rating scale for global mental workload measurement applications, *Proceedings of the Human Factors Society—27th Annual Meeting*, HFES: Santa Monica, CA, pp. 129–133.

Whiting, H. (1969). *Acquiring Ball Skill: A Psychological Interpretation*, Bell: London, U.K.

10 Warning Labels, Instructions, and Product Liability

10.1 LEARNING GOALS

10.2 KEY TOPICS

- Types of product liability
- Hazard classification
- Principles of warning label design

10.3 INTRODUCTION AND BACKGROUND

In addition to promoting safety and healthy work practices in the occupational environment, the field of ergonomics is equally concerned with promoting consumer product safety through safe product design and use (Wilson and Kirk, 1980). During the past several decades, there has been an increasing concern for public safety in the United States in terms of consumer products. The Consumer Product Safety Commission (CPSC), an independent federal regulatory agency, established by the Act of October 27, 1972 (86 Stat. 1207), has the mission to ensure the safety of consumer products manufactured, distributed, and marketed in the American industry. The commission has primary responsibility for establishing mandatory product-safety standards in order to reduce unreasonable risk of injury to consumers from consumer products. It also has the authority to ban hazardous consumer products. The responsibilities are defined as follows:

> The primary responsibility of the CPSC is to protect the public from unreasonable risks of injury that could occur during the use of consumer products. The CPSC also promotes the evaluation of consumer products for potential hazards, establishes uniform safety standards for consumer products, eases conflicting state and local regulations concerned with consumer safety, works to recall hazardous products from the marketplace, and selectively conducts research on potentially hazardous products. (CPSC, 2011)

Additionally, the Consumer Product Safety Act (15 U.S.C. 2051 et seq. [1972]) authorizes the commission to conduct extensive research on consumer product standards, to engage in broad consumer, industry information, and education programs, and to establish a comprehensive injury-information clearinghouse (CPSC website,

2011). This responsibility is increasingly more challenging with the globalization of product development and manufacturing.

Manufacturers, service industries, and recreational environments are continually implementing strategies that promote safer products, environments, and processes to reduce the likelihood for injury to employees, and organizational liability. Ensuring safety in design and redesign is accomplished through

- Engineering design
- Administrative controls
- Hazard communication through warning labels

The distinction between warnings and instructions is offered by Laughery and Wogalter (1997) stating that warnings are communication about safety while instructions may or may not contain information regarding safety.

The least desirable among these three approaches to risk reduction is warning as this approach is completely dependent upon the response of the consumer to be effective. This approach should be used in conjunction with design and administrative controls.

As technology increases, product liability should decrease; however, this has not always been the case in all industries. The improvement in product design and manufacture leads to better products but it also introduces new hazards (i.e., risk of damage to the ears with enhanced quality earphones). Technologies such as computers allow for better design, not only in the product, but also in the workspaces using resources such as simulation to anticipate and model potential hazards to employees and consumers (Ettlie, 1998).

10.3.1 IMPACT OF PRODUCT LIABILITY ON ERGONOMICS AND HUMAN FACTORS IN PRODUCT DESIGN

With the threat of product liability looming over manufacturers, ergonomist often work closely with safety engineers in product design to ensure a well-designed product. U.S. Ergonomics, Inc. (2010), a multidisciplinary team of Certified Professional Ergonomists, "specializes in the application of measurement-based technologies to assess the human response to product and/or workplace design. The results of these analyses are used to improve comfort and efficiency while reducing fatigue potential, injury risk and liabilities". This ensures that the product designed meets the requirements as well as decreases potential liability. This testing is important, because oftentimes manufacturers will protect themselves by deciding not to pursue a product after testing, in order to avoid product liability. This may lead to "discontinued product research, cut backs on introducing new product lines with known risks, and increase product cost in order to design to mitigate risks. Ultimately, the abuse of product liability laws offers consumers fewer domestic products at higher prices and compromises the competitiveness of U.S. firms in foreign and domestic markets" (American Tort Reform Association, 2007). However, the benefits of product liability are many including safe and healthy consumer products as well as responsible manufactures.

10.4 DEFINITIONS

In this section, several terms are defined, particularly the concepts of hazard, danger, and risk perception. These terms are sometimes used in different ways with different meanings. Therefore, it is important to be clear as to their meaning in this context.

Hazard is a term that is used to describe a set of circumstances that can result in injury, illness, or property damage. Such circumstances may include characteristics of the environment, of equipment, and of a task someone is performing. It is important to take into consideration that circumstances can also include characteristics of the people involved, for example, their abilities, limitations, and knowledge. Danger can be defined in several ways. In this chapter, it is defined as the product of hazard and likelihood. Risk perception involves the overall awareness and knowledge regarding the hazards, likelihoods, and potential outcomes of a situation or set of circumstances.

10.5 PRODUCT LIABILITY

According to the Cornell University Law School's Legal Information Institute, product liability is defined as follows (Cornell, 2009):

> Products liability refers to the liability of any or all parties along the chain of manufacture of any product for damage caused by that product. This includes the manufacturer of component parts (at the top of the chain), an assembling manufacturer, the wholesaler, and the retail store owner (at the bottom of the chain). Products containing inherent defects that cause harm to a consumer of the product, or someone to whom the product was loaned, given, etc., are the subjects of products liability suits. While products are generally thought of as tangible personal property, products liability has stretched that definition to include intangibles (gas), naturals (pets), real estate (house), and writings (navigational charts). (Legal Information Institute, 2010)

The theories of product liability are generally classified in one of three categories and have been written about extensively (Gasaway, 2002; Kinzie, 2002; Moore, 2001). These theories are

1. Negligence
2. Strict liability
3. Breach of warranty

Sometimes all three theories are pursued in one case. Also, strict liability and breach of warranty may include cases of misrepresentation. According to the legal dictionary (http://legal-dictionary.thefreedictionary.com/Product+Liability), "Misrepresentation refers to a situation when a manufacturer, distributor or seller of a product gives consumers false security about the safety of the product; this is often done by drawing attention away from the hazards of its use. Fault lies in the intentional concealment of potential hazards or in negligent misrepresentation."

Discussions of each of these theories are provided in the following sections (Black's Law Dictionary, 2011).

10.5.1 NEGLIGENCE

A negligence theory requires the plaintiff to prove four elements:

- First, it must be shown that the defendant owed a duty to the consumer.
- Second, that the manufacturer did not guard against injuries likely to result from a reasonably foreseeable misuse of the product.
- Third, that the manufacturer breached its duty (by applying the aforementioned design defect, manufacturing defect, or failure to warn theories).
- Fourth, plaintiff need also prove he or she was injured and that the manufacturers' breach caused the injury.

Failure to warn may be the case if, a manufacturer who sells a painkiller without warning that use of the drug with alcohol could impair the user's ability to operate machinery. In this case, the manufacturer may be liable if the consumer is injured in this fashion.

Defective design claims that a product is, at the time it is sold, in a defective condition and is unreasonably dangerous to the ordinary consumer. For example, a seller who designs a circular power saw that does not include a blade guard may be liable if the user suffers an injury from the unguarded blade.

10.5.2 STRICT LIABILITY

Strict liability is essentially identical to negligence theory except the plaintiff need not show knowledge or fault on the manufacturer's part. In other words, the plaintiff still must show the four elements of negligence but the knowledge of the product's unreasonable danger (if the product is proven unreasonably dangerous) is attributed to the manufacturer. The following question then arises: "would the reasonable manufacturer, with such knowledge of the product's risk, have produced the product?"

Usually, the defendant (manufacturer) has knowledge or should have had knowledge and the attribution of knowledge is irrelevant. In cases where it cannot fairly be said that the manufacturer had knowledge or should have had knowledge of a product's risk (such as, possibly, asbestos manufacturers many decades ago), if strict liability is properly applied knowledge will be attributed. Strict liability concerns only the condition of the product while negligence is concerned not only with the product but also the manufacturer's conduct. Causation and damages are the same as they are for negligence but the primary difference in strict liability lies with defect. Seven factors must be analyzed to verify a defect (Black's Law Dictionary, 2011):

- The product's usefulness
- The availability of safer products to meet the same need
- The likelihood and probable seriousness of injury
- The obviousness of the danger
- The public expectation of the danger
- The avoidability of injury by care in the use of the product, including the effect of instructions and warnings
- The manufacturer's or seller's ability to eliminate the danger of the product without making it useless or unduly expensive

However, it is important to point out that "strict liability" does not mean "absolute liability." Thus, just because a person is injured, he or she cannot assert strict liability and automatically recover but rather the injured consumer who is asserting strict liability still must prove his or her right to compensation.

Strict liability does not take into account the conduct of the parties but instead focuses on the quality of the product that caused the injury. The consumer only needs to prove that the product is defective. In order to establish strict liability, the consumer needs to prove three main elements: causation, damages, and defect.

10.5.3 Breach of Warranty

The final product liability theory is that of breach of warranty. This theory is based on the premise that "every product comes with an implied warranty that it is safe for its intended use. A defective product that causes injury was not safe for its intended use and thus may constitute a breach of warranty. A manufacturer cannot simply disclaim such a warranty but will be held responsible if its product is deemed defective.

The performance and reliability of any product depends on the effectiveness of the product design and guidance provided to the user. Regardless of the hardware and software performance characteristics of a design, some designs result in better performance, are more reliable and safer than others (Priest and Sanchez, 1988). According or Priest and Sanchez, the best practices to successfully design for people are as follows (Priest and Sanchez, 1988):

- Simple and effective human interface: this is necessary to promote good performance, reduce human errors and system-induced errors. A simple design also increases the number of potential users when the principles of anthropometrics, ergonomics, repairability, safety, and product liability are considered.
- Functional task allocation analysis: this maximizes system performance by effectively dividing performance, control, and maintenance task between the machine, assistive devices, and the operator.
- Task analysis, failure modes, maintenance analysis, safety analysis, and product liability analysis are used to analyze human requirements, identify and correct potential problems, and predict operator performance.
- Design guidelines and analysis promote mistake proofing with simplified, standardized/common tasks and human interface designs to improve personal performance.
- Prototype testing of task performance including repair and manufacturing is used to identify design improvements, potential human errors, and hazards.
- Effective design documentation is critical for user, manufacturing, and repair instructions.
- Safety hazard and product liability analyses identify and correct all potential hazards including all foreseeable uses, foreseeable misuses, and modifications.

Safety and product liability analyses require that a critical review of a product be performed in appropriate and inappropriate product use. Safety engineers and ergonomist must consider foreseeable misuse of the product. This should result in the evaluation or consideration of all possible hazards. The implementation of hazard, ergonomic, and safety analysis in the design stage are very valuable in the design phase; however, the greatest dividends of these actions are in the reduction in product-related accidents, product misuses, recalls, and liability costs.

10.5.3.1 What Is a Hazard?

A hazard is any condition or situation (existing or potential) that is capable of injuring people or damaging the product, adjacent property, or the environment (Priest and Sanchez, 1988). The steps in an effective hazard analysis are as follows (adapted from Priest and Sanchez, 1988):

- Identify hazards by level.
- Identify the reasons and factors that can produce the hazard.
- Evaluate and identify all potential effects of the hazard.
- Categorize the identified hazards as catastrophic, critical, marginal, or negligible.
- Implement design changes that minimize the number and level of hazards.
- Determine if administrative controls (i.e., training, personal protective equipment) should be used.
- Develop a plan to provide warnings and instructions to end users to inform of minimal hazards (or any other hazards) that are not removed from the design.

DoD MIL STD 882, System Safety Program Requirements, classifies hazard severity as follows:

It is very important for manufacturers to evaluate the hazardous characteristics of their products during the product development process. This assessment can be part of a preliminary hazard analysis, a failure mode and effects analysis, or some other test of a product before its final design is approved and further actions to eliminate or control any hazards take place (shown in Table 10.1).

10.5.4 Warnings

Warnings inform people of hazards and provide instructions as to how to deal with them in order to avoid or diminish undesirable consequences. Warnings are used to address environmental hazards as well as hazards associated with the use of products. The increased use of warning signs has been accompanied by regulations,

TABLE 10.1
DoD Hazard Classification

Severity	Category	Accident Definition
Catastrophic	I	Death or total system loss
Critical	II	Severe injury, severe occupational illness, or major system damage
Marginal	III	Minor injury, minor occupational illness, or minor system damage
Negligible	IV	Less than minor injury, occupational illness, or system damage

standards, and guidelines as to when and how to warn. Human factors specialists, or ergonomists, have played a significant role in the research of warnings and the technical literature that has resulted.

Warnings are intended to improve safety by decreasing accidents or incidents that result in injury, illness, or property damage. At another level, warnings are intended to influence or modify people's behavior in ways that improve safety. At still another level, warnings are meant to provide information that enables people to understand hazards, consequences, and appropriate behaviors, which in turn allows them to make informed decisions. Also, warnings can be viewed as communications. In this context, it is important to describe the typical communications model or theory, because it can have implications in the design and implementation of warnings. A typical and basic model includes a sender, a receiver, a channel or medium through which a message is transmitted, and the message. The receiver can be the user of the product, the worker, or any other person to whom the safety information must be transmitted. The medium refers to the channels or routes through which information is transmitted. Additional factors impacting the quality of communication are the design of the message, environment, and external influences. Some of these factors are shown in Table 10.2 (Eastman Kodak Company, 1986).

The sender of a message must design written material so that it is comprehensible, legible, and readable (column 1). Factors that influence each of these design

TABLE 10.2
Factors Affecting the Written Communication

Design of Message → by Sender	Factors Affecting → Message Transmission	Elements Influencing Receipt of the Message
Comprehensibility	*Environment*	*Discrimination*
• Purpose	• Viewing distance	• Visual abilities
• User knowledge	• Viewing angle	
• Brevity	• Illumination	*Interpretation*
• Accuracy	• Deterioration	• Language skills
• Clarity	• Competing displays	• Situation knowledge
	• Timing pressure	
Legibility		*Recall*
• Font style		• Time delay
• Font size		• Interference
• Colors		
Readability		
• Borders		
• Layout		
• Abbreviations		
• Spacing		
• Case		

Source: Adapted from Caplan, S.H., 1975; Eastman Kodak Company, 1986.

objectives are listed. As the message is transmitted through the environment (column 2), other factors may affect how well it can be picked up by the receiver. The receiver's characteristics (column 3) also influence the accuracy with which information is communicated. Attention to the factors listed in each part of the process should reduce the potential for errors in written communication.

Ideally, all products should be designed with safety as a priority. However, some products have inherent risks and safety is not always achieved through design alone. Three approaches when making a product safe for consumer use are as follows:

- Firstly, the dangerous features out of the product should be designed. If that is not possible, the second approach is to mitigate or protect from the hazard or engineered out of the design (i.e., guarding or shielding).
- Secondly, administrative controls or personal protective equipment can be used to mitigate risk in some cases. This includes requirements for training, certification on product use, and provision of protective items such as work goggles.
- Finally, if a hazard is still evident, adequate warnings and instructions must be provided for proper use and reasonable foreseeable misuse. Warning signs or labels are used to warn users of the dangers that come with the use of the product. Although, instructions for products often contain warnings, warnings are not instructions. Instructions are provided for the product's safe use and to prevent any injuries or damages.

The objective of a warning sign is to alter behavior. The warning must be sensed audibly, visually, or through olfactory (i.e., smell of natural gas), received through reading or listening, and understood, which should ultimately lead to adherence. Warning design can be a very complex task and where the risks of serious injury or death are involved, warnings should be tested for effectiveness with potential users. Several principles or rules are used to guide when a warning should be used. The following list can be used to determine whether a warning is necessary (Laughery and Wogalter, 1997):

1. A significant hazard exists.
2. The hazard, consequences, and appropriate safe modes of behavior are not known by the people exposed to the hazard.
3. The hazards are not open and obvious; that is, the appearance and function of the environment or product do not communicate them.
4. A reminder is needed to assure awareness of the hazard at the proper time. This concern is especially important in situations of high task loading or potential distractions.

If a warning label is needed, it is important to design and place it such that it will be perceived, sensed and received. In order to determine if a warning will be easily sensed, the following factors need to be considered:

- Orientation
- Location

- Size, including font size
- Shape
- Color
- Graphical design
- Contrast
- Attention getting aspects, for example, a flashing light

All warning signs or labels should follow the guidelines as outlined by the American National Standard for Product Safety: Signs and Labels (ANSI Z535.4-1991), which requires all warnings to comply with the following five criteria (Figure 10.1):

1. The signal word is appropriate to the level of hazard.
   a. Danger
   b. Warning
   c. Caution
2. The statement describes the hazard.
3. The probable consequences of involvement with the hazard are described.
4. Instructions on how to avoid the hazard are included.
5. The label has the appropriate colors, graphics, and pictorials.
   a. Danger
      i. White letters, red background
   b. Warning
      i. Black letters, orange background
   c. Caution
      i. Black letters, yellow background

FIGURE 10.1   Example of Warning label.

The levels of signal words are as follows (Figure 10.2):

- Danger
  - Used when there is an immediate hazard and if encountered it can result in severe injury or death
- Warning
  - Used for hazards or unsafe products
- Caution
  - For hazards or unsafe practices that could usually result in minor personal injury, product damage, or property damage

Sensing a warning does not necessarily mean that it will be received. Several factors can influence receipt or reading of warnings, such as familiarity with the product, complexity of warning, length of warning, and numerous others. It is important that the words and graphics of a warning are understood by the user population. The minimum information necessary in a warning includes the following:

- Signal word
  - To convey the gravity or severity of the risk
- Hazard
  - Description of the nature of the hazard
- Consequences
  - Describes what will happen in case the warning is not heeded
- Instructions
  - Provides the appropriate behavior needed to reduce or eliminate the hazard

Sensing, receiving, and understanding a warning does not necessarily mean that it will also be heeded. Cost of compliance is one of the factors that can impact a

FIGURE 10.2   Samples of signal word usage.

person's likelihood to heed a warning. Cost of compliance refers to the degree of energy, time, resources, or deviation from the activity at hand that complying with the warning requires. The high cost of compliance results in a low probability of compliance. Other factors affecting a person's decision to follow a warning include the risk acceptance level of the individual and the ability to remember the warning.

10.5.5 PRIORITIZING WARNINGS

An important concern with hazards deals with the question of what hazards to warn about when multiple hazards exist. How are priorities defined when deciding what to include or omit, how to sequence them, or how much emphasis to give them? When a hazard is already known and understood, obvious warnings may not be needed however, depending on the user and environment, the designer may still want to provide a warning. Other factors that should be considered include (Laughery and Wogalter, 1997):

- Likelihood
  - The more likely an undesirable event is to occur, the greater the priority that it should be warned.
- Severity
  - The more severe the potential consequences of a hazard, the greater priority that it should be warned. If a chemical product poses a skin contact hazard, a higher priority would be given to a severe chemical burn consequence than if it were a minor rash.
- Practicality
  - There are occasions when limited space, such as a small label or limited time, for example, a television commercial, does not permit all hazards to be addressed in a single component of the warning system.
  - As a general rule, unknown hazards leading to more severe consequences and/or those more likely to occur would have priority for the primary warning component, such as on the product label, whereas those hazards with lower priority would be addressed in other warning components, such as package inserts or manuals.

10.5.6 ACTIVE VERSUS PASSIVE WARNINGS

Active warnings usually have sensors that detect inappropriate use or alert the user to an imminent danger. Although active warnings are preferred to passive warnings, passive warnings are used more commonly. Passive warnings can be found on the label of a product as well as in the instructions within the users' manual of the product.

10.5.7 DESIGN OF WARNINGS

There are several principles that should be considered when taking receiver characteristics into account during the design of warnings (Woolgather and Laughery, 2006):

- Principle 1
  - Know the receiver. Gathering information and data about relevant receiver characteristics may require time, effort, and money, but without it the warning designer and ultimately the receiver will be at a serious disadvantage.
- Principle 2
  - When variability exists in the target audience, design warnings for the low-end extreme.
  - Do not design for the average.
- Principle 3
  - When the target audience consists of subgroups that differ in relevant characteristics and efforts to apply that knowledge, warnings generally should be market tested.
  - Such tests may consist of trying it out on a target audience sample to assess comprehension and behavioral intentions.

Labels and signs for warnings must be carefully designed by adhering to the most recent government laws and regulations, recognized national and international standards, and the best ergonomic applications. It is important to carefully analyze the product to determine if any hazards associated with the product exist as well as anticipate both the intended and unintended uses of the product. The design of warnings can and should be viewed as an integral part of the design system. Warnings cannot and should not be expected to serve as a cure for bad design. Generally, the strategy that is used to produce a safe product involves the following steps:

- Remove the hazard.
- Limit access to the hazard.
- Inform the user of the hazard by using labeling and instruction manuals.
- Train the user to avoid the hazard.

If satisfaction is not achieved either through the elimination or minimization of the hazard, the remaining choice is to either design to mitigate the hazard or completely ban the product, which requires federal government involvement. All warning labels should accomplish three things:

1. Get the user's attention.
2. Describe the danger in clear terms.
3. Give specific instructions on how to avoid any injury from using the product.

Typically, seven steps are used when designing warning signs or labels and they are summarized in the following:

1. Specifically define
   a. The risk or hazard
   b. The consequences of exposure to the hazard
   c. The audience or receiver of the message
   d. How to avoid the hazard

2. Determine and evaluate the best method to communicate the message.
3. Draft the wording.
4. Evaluate the wording by using a communication specialist and legal counsel.
5. Perform user testing, evaluating results with established pass-fail criteria.
   a. Was the message noticed?
   b. Was the message understood?
   c. Was avoidance behavior correctly carried out by the users?
6. Revise the warning based on the results of user testing.
7. Retest the warning if it is necessary. Finally, print and distribute.

There are eight criteria that are quite useful in the design and assessment of warnings and they include the following:

- Attention
  - Warnings should be designed so as to attract attention.
- Hazard information
  - Warnings should contain information about the potential outcomes.
- Consequence information
  - Warnings should contain information about the potential outcomes.
- Instructions
  - Warnings should instruct about appropriate and inappropriate behaviors.
- Comprehension
  - Warnings should be understood by the target audience.
- Motivation
  - Warnings should motivate people to comply.
- Brevity
  - Warnings should be as brief as possible.
- Durability
  - Warnings should last and be available as long as needed.

There are several dimensions of receiver competence that may be relevant to the design of warnings. For example, sensory deficits might be a factor in the ability of some special target audience to be directly influenced by a warning. A blind person would not be able to receive a written warning nor would a deaf person be able to receive an auditory warning. There are three characteristics of receivers that relate to cognitive competence and are important in the design of warnings:

- Technical knowledge
  - The communication of hazards associated with medications, chemicals, and mechanical devices.
  - If the target audience does not have technical competence, the warning may not be successful.
- Language
  - There are several subgroups in the American society who speak and read languages other than English. As trade becomes more international,

requirements for warnings to be directed to non-English readers will increase.
- Solutions to this problem include designing warnings that are stated in multiple languages and the use of pictorials or graphics.
- Reading ability
  - The general recommendation for common target audiences is that the reading level be in the grade 4–6 range.
  - An important point on reading ability deals with illiteracy. There are about 16 million functionally illiterate adults in the American population. Therefore, successful communication of warnings may require more than simply keeping reading levels to a minimum.
  - Use pictorials or graphics to supplement text.

10.6   SUMMARY

As technology shortens the product development life cycle, it will become increasingly more important to integrate ergonomics in product development, testing, and safety communication. The principles of warning and communication of product information should be adhered to throughout product development as the application of human factors and ergonomics knowledge in the design of consumer products, warnings, and instructions has proven useful in reducing risk of incidents.

Case Study

FDA Orders Easy-to-Understand Warnings for Prescription Drugs

February 9, 2004

By *Jeanie Croasmun*

In an attempt to make prescription drug warnings more readily understandable by consumers, decision-makers at the FDA last week ordered prescription drug manufacturers to start following new guidelines that should make the fine-print of prescription ads more consumer-friendly. In other words, warning labels should be more ergonomic.

Honing in on the fine print of prescription-drug print ads and the difficulties consumers may have reading and understanding any associated warnings, the FDA's new guidelines recommend the following:

- Lay terms rather than medical jargon in warnings so a potential user of the prescription drug can comprehend the list of possible side-effects.
- A reorganization of ads to put the most common risks and potentially lifesaving information at the beginning of the ad's warning statements.

- Bigger font sizes for warnings, thereby improving readability.
- Relocation of the warnings and side-effects to the same page as the ad rather than behind it.

The motivation behind the new guidelines are to create something patients will both read and understand, something the FDA believes is not presently the case. Patients do not read the lengthy warnings, the FDA said, and when they try, the guidance given is often not understood.

The FDA also believes there may a positive side-effect for drug manufacturers—smaller (and less expensive) ads. "This may be a case where less is more in terms of consumer understanding," FDA commissioner, Mark McClellan, told CNN.com, explaining that a smaller but more precisely-written ad may have a greater impact on the reader than just printing an all-encompassing warning statement.

Critics of the guidelines, however, believe that this may also open up the door to warnings that are too vague or misleading. Representative Henry Waxman, D-California, told CNN that the new warning guidelines could mean that a statement previously reading "may cause acute liver failure," could now be replaced with "can affect liver function."

While providing understandable instructions for use of a product is a key component of ergonomics, warning labels are still a regular source of debate in regard to user understandability. Recently, for example, the National Highway Traffic Safety Administration (NHTSA) scrapped plans to implement a standard set of dashboard warning lights and symbols for auto makers after it was determined that in tests, consumers weren't able to connect the proper meaning to the warnings. Ironically, the NHTSA's goal in creating the new set of warning lights was to develop a more easily understandable system for consumers.

Sources: Reprinted directly as shown with permission from Ergoweb, 2011; CNN.com; Ergonomics Today™

EXERCISES

10.1 Explain the four types of product liability classifications from a legal perspective.
10.2 What have been some of the historic benefits in manufacturing as a result of product liability cases?
10.3 Identify the branch of the government in your country (or the United States) responsible for consumer product safety and go to the website and obtain an overview of the process used to establish consumer product-safety guidelines.
10.4 Discuss the types of keywords that are used in warning design and give an example where each should be used in warning design.
10.5 Design a warning label for a physically intensive computer-based game (i.e., Wii game) that is used for recreational use.

REFERENCES

American Tort Reform Association. (2007). http://www.atra.org/ (accessed March 12, 2011).

Caplan, S.H. (1975). In Eastman Kodak, (1986). *Ergonomic Design for People at Work — Volume 2*, Van Nostrand Reinhold Company Inc.: New York.

Consumer Product Safety Commission is back on track, *Manufacturing News*, January 12, 2000.

Cornell website. (2009). http://www.law.cornell.edu/wex/Products_liability

Ergoweb website. (2011) http://www.ergoweb.com/resources/casestudies/

Ettlie, J.E. (1998). R&D and global manufacturing performance. *Manage. Sci.*, 44, 1–11.

Gasaway, R.R. (2002). The problem of tort reform: Federalism and the regulation of lawyers. *Harv. J. Law Public Policy*, 25(3), 953(11).

Gooden, R. Reduce the potential impact of product liability on your organization. *Quality Progress*, January 1995.

Heafey, R. and Kennedy, D. (2006). *Product Liability: Winning Strategies and Techniques*, Law Journal Press, ALM Media: New York.

Kinzie, M.A. (2002). *Product Liability Litigation*, West/Thomson Learning: Albany, NY.

Laughery, K. and Wogalter, M. (1997). Warnings and risk perception, In: *Handbook of Human Factors and Ergonomics*, John Wiley & Sons: New York, Chap. 36, pp. 1174–1198.

Legal Information Institute. (2010) http://www.law.cornell.edu/

Martin, B.J., and S.W. Deppa. (1997). Human factors in the revised ANSI Z535. 4 Standard for safety labels. Paper presented at *Human Factors and Ergonomics Society Annual Meeting Proceedings*, pp. 821–825.

Moore, M.J. (2001). *Product Liability Entering The Twenty-First Century: The U.S. Perspective*, AEI-Brookings Joint Center for Regulatory Studies: Washington, DC.

Mulherin, J. (2001). Geier v. American Honda Motor Company, Inc.: Has the Supreme Court extended the pre-emption doctrine too far? *J. Natl Assoc. Adm. Law J.*, 21.

Plog, B.A., Quinlan, P., and National Safety Council. (2002). *Fundamentals of Industrial Hygiene*, National Safety Council: Chicago, IL.

Postrel, V. When you're in the danger business. *Forbes*, January 25, 1999.

Sanchez, J.M., and Priest, J.W. (1988). Engineering Design for Producibility and Reliability, Dekker; New York.

Siomkos, G. and Shrivastava, P. (1993, October). Responding to product liability crises. *Long Range Plann.*, 26(5), 72–79.

U.S. Ergonomics, Inc. (2010). http://www.us-ergo.com/current_events/index.asp. (accessed November 29, 2010).

Wilson, J.R. and Kirk, N.S. (1980). Ergonomics and product liability. Applied Ergonomics, 11(3), 130–136.

Index

A

AADMS, *see* Automated anthropometric data measurement system
Acetylcholine (ACh), 102, 106
Actin
    muscular contractile system, 100, 101
    sliding filament model, 101
ADA, *see* Americans with Disabilities Act
Adenosine diphosphate (ADP), 106
Adenosine triphosphate (ATP), 101
American National Standards Institute (ANSI), 13, 16
Americans with Disabilities Act (ADA), 124
AnthroKids, 144
Anthropometric design
    aids and data sources
        AnthroKids, 144
        CAESAR, 143–144
        International Anthropometric Resource, 144
        International Journal of Clothing Science and Technology, 144–145
        U.S. Army Anthropometry Survey, 144
    biomechanics and application, *see* Biomechanics
    data collection, 145–146
    gender, ethnicity and age measurement, 122
    goals, 121
    human body and parts, measurement, 121–122
    measurements
        physical properties, body segments, 134
        seated, 134
        stature, 133–134
    OSHA website
        solution, 155–156
        workstation layouts, 155
    people, jobs, 121
    principles
        guidelines, 123–124, 125
        myth, average human, 127
        universal design, 124–127
    sophisticated method, 122–123
    static and dynamic dimensions, 134–137, 138–140
    structural/static, 123
    terminology
        body measures and applications, 129–132
        measurements, 128
        planes, 128, 132–133

    tools
        automated technology, 143
        hand measurement instruments, 141–142
        HSIIAC and CSEIAC, 143
        instruments, 140
        manual, 141
        photographs, 142
        scanner and laser, 142
        special calipers, 140
        spreading calipers, 140
    use, 122
Anthropometry, ergonomics
    interdisciplinary nature, 5
    problems, 5
Applications, ergonomics
    approaches, 6–7
    preparation, 19–20
    and research
        Department of the Interior, 9
        DoD, 9
        FAA, 10
        Federal Highway Administration, 9
        NASA, 9
        National Highway Traffic Safety Administration, 9
        National Institute of Standards and Technology, 9
        NIOSH, 10
        NRC, 10
        OSHA, 10
ASCC, *see* Australian Safety and Compensation Council
ATP, *see* Adenosine triphosphate
Attention
    definition, 286
    divided, 287
    selective and focused, 287
    sustained, 286
Audition
    ear, 61
    iPods and MP3(4) impact
        hearing loss, 65–66
        higher sound levels, 64
    NIHL, 62, 64, 65
    noise effects, 63–64
    ossicles, 61
    speech intensity levels, 62
Australian Safety and Compensation Council (ASCC), 11

Printed in the United States
by Baker & Taylor Publisher Services